AF347753

Twentieth Century Inventions

A Forecast

by

George Sutherland

Twentieth Century Inventions
A Forecast
by George Sutherland

Copyright © 2024

All Rights reserved.

No part of this publication may be reproduced, stored in a retrieval system, or transmitted in any form or by any means, electronic, mechanical, photocopying or Otherwise, without the written permission of the publisher.
The author/editor asserts the moral right to be identified as the author/editor of this work.

ISBN: 978-93-68099-66-6

Published by

DOUBLE 9 BOOKS

2/13-B, Ansari Road
Daryaganj, New Delhi – 110002
info@double9books.com
www.double9books.com
Tel. 011-40042856

This book is under public domain

ABOUT THE AUTHOR

George Sutherland was an insightful thinker and writer, best known for his work on technological advancements and their potential societal impact. In his book Twentieth Century Inventions: A Forecast, Sutherland explored a wide range of innovations that he predicted would transform the world. One of his most forward-thinking ideas was the use of wireless telegraphy to synchronize clocks across a given area, ensuring that all timepieces within that region would be set to a single, standard time. This concept, far ahead of its time, anticipated the global systems of time synchronization that would later be made possible by technologies like GPS and atomic clocks. Sutherland's work reflected his deep interest in how emerging technologies could streamline daily life and improve efficiency. His writing not only captured the excitement of a rapidly changing era but also demonstrated his ability to envision the long-term implications of these innovations. Through his predictions, Sutherland contributed to the broader dialogue about the transformative potential of science and technology during the early 20th century.

CONTENTS

PREFACE...7

CHAPTER I
INVENTIVE PROGRESS 11

CHAPTER II
NATURAL POWER 20

CHAPTER III
STORAGE OF POWER 34

CHAPTER IV
ARTIFICIAL POWER......................................43

CHAPTER V
ROAD AND RAIL 52

CHAPTER VI
SHIPS 66

CHAPTER VII
AGRICULTURE...76

CHAPTER VIII
MINING...86

CHAPTER IX
DOMESTIC..99

CHAPTER X
ELECTRIC MESSAGES, ETC.........................109

CHAPTER XI
WARFARE...117

CHAPTER XII
MUSIC...125

CHAPTER XIII
    ART AND NEWS 132

CHAPTER XIV
    INVENTION AND COLLECTIVISM 138

PREFACE

Twenty years ago the author started a career in technological journalism by writing descriptions of what he regarded as the most promising inventions which had been displayed in international exhibitions then recently held. From that time until the present it has been his constant duty and practice to take note of the advance of inventive science as applied to industrial improvement—to watch it as an organic growth, not only from a philosophical, but also from a practical, point of view. The advance towards the actual adoption of any great industrial invention is generally a more or less collective movement; and, in the course of a practice such as that referred to, the habit of watching the signs of progress has been naturally acquired.

Moreover, it has always been necessary to take a comprehensive, rather than a minute or detailed, view of the progress of the great industrial army of nineteenth century civilisation towards certain objectives. It is better, for some purposes of technological journalism, to be attached to the staff than to march with any individual company—for the war correspondent must ever place himself in a position from which a bird's-eye view is possible. The personal aspect of the campaign becomes merged in that which regards the army as an organic unit.

It may, therefore, be claimed that, in some moderate degree, the author is fitted by training and opportunities for undertaking the necessarily difficult task of foretelling the trend of invention and industrial improvement during the twentieth century. He must, of course, expect to be wrong in a certain proportion of his prognostications; but, like the meteorologists, he will be content if in a fair percentage of his forecasts it should be admitted that he has reasoned correctly according to the available data.

The questions to be answered in an inquiry as to the chances of failure or success which lie before any invention or proposed improvement are, first, whether it is really wanted; and, secondly, whether the environment in the midst of which it must make its début is favourable. These requirements generally depend upon matters which, to a large extent, stand apart from the personal qualifications of any individual inventor.

In the course of a search through the vast accumulations of the patent specifications of various countries, the thought is almost irresistibly forced upon the mind of the investigator that "there is nothing new under the sun". No matter how far back he may push his inquiry in attempting to unveil the true source of any important idea, he will always find at some antecedent date the germ, either of the same inventive conception, or of something which is hardly distinguishable from it. The habit of research into the origin of improved industrial method must therefore help to strengthen the impression of the importance of gradual growth, and of general tendencies, as being the prime factors in promoting social advancement through the success of invention.

The same habit will also generally have the effect of rendering the searcher more diffident in any claims which he may entertain as to the originality of his own ideas. Inventive thought has been so enormously stimulated during the past two or three generations, that the public recognition of a want invariably sets thousands of minds thinking about the possible methods of ministering to it.

Startling illustrations of this fact are continually cropping up in the experiences of patent agents and others who are engaged in technological work and its literature. The average inventor is almost always inclined to imagine — when he finds another man working in exactly the same groove as himself — that by some means his ideas have leaked out, and have been pirated. But those who have studied invention, as a social and industrial force, know that nothing is more common than to find two or more inventors making entirely independent progress in the same direction.

For example, while this book was in course of preparation the author wrote out an account of an application of wireless telegraphy to the purpose of keeping all the clocks within a given area correct to one standard time. Within a few days there came to hand a copy of *Engineering* in which exactly the same suggestion was put forward, and an announcement was made to the effect that Mr. Richard Kerr, F.G.S., had been working independently on the same lines, the details of his method of applying the Hertzian waves to the purpose being practically the same as those sketched out by the author. This is only one of several instances of coincidences in independent work which have been noticed during the period while this volume was in course of preparation.

It may, therefore, be readily understood that the author would hardly like to undertake the task of attempting to discriminate between those forecasts in the subsequent pages which are the results of his own original suggestions, and those which have been derived from other sources.

Whatever is of value has in all probability been thought of, or perhaps patented and otherwise publicly suggested, before. At any rate, the great majority of the forecasts are based on actual records of the trials of inventions which distinctly have a future lying before them in the years of the twentieth century.

In declining to enter into questions relating to the original authorship of the improvements or discoveries discussed, it should not be supposed that any wish is implied to detract from the merits of inventors and promoters of inventions, either individually or collectively. Many of these are the heroes and statesmen of that great nation which is gradually coming to be recognised as a true entity under the name of Civilisation. Their life's work is to elevate humanity, and if mankind paid more attention to them, and to what they are thinking and doing, instead of setting so much store by the veriest tittle-tattle of what is called political life, it would make much faster progress.

Some of the industrial improvements referred to in the succeeding pages are necessarily sketched in an indefinite manner. The outlines, as it were, have been only roughed in; and no attempt has been made to supply particulars, which in fact would be out of place in an essay towards a comprehensive survey in so small a space. It is upon the wise and skilful arrangement of details that sound and commercially profitable patents are usually founded, rather than upon the broad general principles of a proposed industrial advance or reform.

During the twentieth century this latter fact, already well recognised by experts in what is known as industrial property, will doubtless force itself more and more upon the attention of inventors. Every specification will require to be drawn up with the very greatest care in observing the truth taught by the fable of the boy and the jar of nuts. So rapidly does the mass of bygone patent records accumulate, that almost any kind of claim based upon very wide foundations will be found to have trenched upon ground already in some degree taken up.

Probably there is hardly anything indicated in this work which is not—in the strict sense of the rules laid down for examiners in those countries which make search as to originality—common public property. The labour involved in gathering the data for a forecast of the inventions likely to produce important effects during the twentieth century has been chiefly that of selecting from out of a vast mass of heterogeneous ideas those which give promise of springing up amidst favourable conditions and of growing to large proportions and bearing valuable fruit. Such ideas, when planted in the soil of the collective mind through the medium of official or other records, generally require for their germination a longer time than that for which the patent laws grant protection for industrial property. Many of them, indeed, have formed the subjects of patents which, from one reason or another, lapsed long before the expiration of the maximum terms. Nature is ever prodigal of seeds and of "seed-thoughts" but comparatively niggardly of places in which the young plant will find exactly the kind of soil, air, rain, and sunshine which the young plant needs.

If any one requires proof of this statement he will find ample evidence in support of it in the tenth chapter of Smiles's work on *Industrial Biography*, where facts and dates are adduced to show that steam locomotion, reaping machines, balloons, gunpowder, macadamised roads, coal gas, photography, anæsthesia, and even telegraphy are inventions which, so far as concerns the germ idea on which their success has been based, are of very much older origin than the world generally supposes. The author, therefore, submits that he is justified in referring inventions to the century in which they produce successful results, not to that in which they may have been first vaguely thought of. And in this view it is obvious that many of those patents and suggestions which have been published in current literature during the nineteenth century, but which, although pregnant with mighty industrial influences, have not yet reached fruition, are essentially inventions of the twentieth century. More than this, it is extremely probable that the great majority of those ideas which will move the industrial world during the next ensuing hundred years have already been indicated, more or less clearly, by the inventive thought of the nineteenth century.

George Sutherland.

December, 1900.

CHAPTER I
INVENTIVE PROGRESS

The year 1801, the first of the nineteenth century, was *annus mirabilis* in the industrial history of mankind. It was in that year that the railway locomotive was invented by Richard Trevithick, who had studied the steam engine under a friend and assistant of James Watt. His patent, which was secured during the ensuing year, makes distinct mention of the use of his locomotive driven by steam upon tramways; and in 1803 he actually had an engine running on the Pen-y-Darran mining tramway in Cornwall. From that small beginning has grown a system of railway communication which has brought the farthest inland regions of mighty continents within easy reach of the seaboard and of the world's great markets; which has made social and friendly intercourse possible in millions of homes which otherwise would have been almost destitute of it; which has been the means of spreading a knowledge of literature, science and religion over the face of the civilised world; and which, at the present moment, constitutes the outward and visible sign of the difference between Western civilisation and that of the Asiatic, as seen in China.

In another corner of the globe, during the year 1801, Volta was constructing his first apparatus demonstrating the material and physical nature of those mysterious electric currents which his friend Professor Galvani of Bologna, who died just two years earlier, had at first ascribed to a physiological source. The researches of the latter, it will be remembered, were begun in an observation of the way in which the legs of a dead frog twitched under certain conditions. The voltaic pile was the first electric battery, and, therefore, the parent of the existing marvellous telegraphic and telephonic systems, while less immediately it led to the development of the dynamo and its work in electric lighting and traction. It brought into harmony much fragmentary knowledge which had lain disjointed in the armoury of the physicist since Dufay in France and Franklin in America had investigated their theories of positive and negative frictional electricities, and had connected them with the flash of lightning as seen in Nature. Thus it became a fresh starting point both for industry and for science.

At the Exposition of National Industry, held in Paris during the year 1801, a working model of the Jacquard loom was exhibited—the prototype of those remarkable pieces of mechanism by which the most elaborately figured designs are worked upon fabrics during the process of weaving by means of sets of perforated cardboards. This was the crowning achievement of the inventions relating to textile fabrics, which had rendered the latter half of the eighteenth century so noteworthy in an industrial sense. It brought artistic designs in articles of common use within the reach of even poor people, and has been the means of unconsciously improving the public taste, in matters of applied art, more rapidly than could have been accomplished by an army of trained artists. The riots in which the mob nearly drowned Jacquard at Lyons for attempting to set up some of his looms were very nearly a counterpart of those which had occurred in England in connection with the introduction of spinning, weaving and knitting machinery.

In Paris, during the first year of the nineteenth century, Robert Fulton, an American, and friend of the United States representative in France, was making trials on the Seine with his first steam-boat—a little vessel imitated by him later on in the first successful steamers which plied on the river Hudson, carrying passengers from New York. At the same time, William Symongton launched the *Charlotte Dundas*, the steam tug-boat which, on the Scottish canals, did the first actually useful work in the conveyance of goods by steam power on the water. These small experiments have initiated a movement in maritime transport which is fully comparable to that brought about on land by the invention of the railway locomotive.

Again, in 1801, Sir Humphry Davy gave his first lecture at the Royal Institution in London, where he had just been installed as a professor, and began that long series of investigations into the chemistry of common things which, taken up by his successor Faraday, gave to the United Kingdom the first start in some of those industries depending upon a knowledge of organic chemistry and the use of certain essential oils.

Public attention at the beginning of the nineteenth century, however, was directed anywhere but towards these small commencements of mighty forces which were to revolutionise the industrial world, and through it also the social and political. If in those days Cornwall was ever referred to, it was not by any means in connection with Trevithick and his steam-engine which would run on rails, but by way of reference to the relations of the Prince of Wales to the Duchy, and the proportion of its revenues which belonged to him from birth.

Glancing over the pages of any history compiled in the early half of the century, the eye will trace hardly the barest allusions to forces, the discoveries in which were, in the year 1801, still in the incipient stage. Canon Hughes, for instance, in his continuation of the histories of Hume and Smollett, devoted some forty pages to the record of that year. The space which he could spare from the demands made upon his attention by the wars in Spain and Egypt, and the naval conflict with France, was mainly occupied with such matters as the election of the Rev. Horne Tooke for Old Sarum, and the burning question as to whether that gentleman had not rendered himself permanently ineligible for Parliamentary honours through taking Holy Orders, and with a miscellaneous mass of topics relating to the merely evanescent politics of the day.

The whole of the effects of invention and discovery in making history during the first year of the century were dismissed by this writer with a casual reference to the augmentation of the productive power of the labouring population through the use of machinery, and a footnote stating that "this was more particularly the case in the cotton manufacture".

Time corrects the historical perspective of the past, but it does not very materially alter the power of the historical vision to adjust itself to an examination of the present day forces which are likely to grow to importance in the making of future history. When we ask what are the inventions and discoveries which are really destined to grow from seeds of the nineteenth into trees of the twentieth century, we are at once confronted with the same kind of difficulty which would present itself to one who, standing in the midst of an ancient forest, should be requested to indicate in what spots the wide-spreading giants of the next generation of trees might be expected to grow. The company promoter labels those inventions in which he is commercially interested as the affairs which will grow to huge dimensions in the future; while the man of scientific or mechanical bent is very apt to predict a mighty future only for achievements which strike him as being peculiarly brilliant.

Patent experts, on the other hand, when asked by their clients to state candidly what class of inventions may be relied upon to bring the most certain returns, generally reply that "big money usually comes from small patents". In other words, an invention embodying some comparatively trivial, but yet really serviceable, improvement on a very widely used type of machine; or a little bit of apparatus which in some small degree facilitates some well known process; or a fashionable toy or puzzle likely to have a good run for a season or two, and then a moderate sale for a few years longer; these are the things to be recommended to an inventor whose main object is to make money. Thus the most qualified experts in patent law and practice do not fail to disclose this fact to those who seek their professional advice in a money-making spirit, as the great majority of inventors do.

The full term of fourteen years in the United Kingdom, or seventeen in the United States, may be a ridiculously long period for which to grant a monopoly to the inventor of some ephemeral toy, although absolutely inadequate to secure the just reward for one who labours for many years to perfect an epoch-making invention, and then to introduce it to the public in the face of all the opposition from vested interests which such inventions almost invariably meet.

Thus the fact that a man has made money out of one class of patents may not be any safe guide at all to arriving at a due estimate of his ideas on industrial improvements of greater "pith and moment," but, on the contrary, it is generally exactly the reverse. The law offers an immense premium for such inventions as are readily introduced, and the inventor who has made it his business to take advantage of this fact is usually one of the last men from whom to get a trustworthy opinion on patents of a different class.

Of the patents taken out during the latter portion of the nineteenth century, many undoubtedly contain the germs of great ideas, and, nevertheless, have excited comparatively little attention from business men or from the general public. It was so in the latter part of the eighteenth century, and history is only repeating itself when the seeds of twentieth century industrial movements are permitted to germinate unseen.

For all practical purposes each invention must be referred to the age in which it actually does useful work in the service of mankind. Thus, Hero of Alexandria, in the third century B.C., devised a water fountain worked by the expansive power of steam. From time to time during the succeeding twenty centuries similar pieces of apparatus excited the curiosity of the inquisitive and the interest of the learned. The clever and eccentric Marquis of Worcester, in his little book published in 1663, *A Century of the Names and Scantlings of Inventions*, generally known as the *Century of Inventions*, gave an account of one application of the power of steam to lift water which he had worked out, probably on a scale large enough to have become of practical service. Thomas Savery and Denis Papin, both of them men of high attainments and great ingenuity, made important improvements before the end of the seventeenth century.

Yet, if we refer to the question as to the proper age to which the steam-engine as a useful invention is to be assigned, we shall unhesitatingly speak of it as an eighteenth century invention, and this notwithstanding the fact that Savery's patent for the first pumping engine which came into practical use was dated 1698. The real introduction of steam as a factor in man's daily work was effected later on, partly by Savery himself and partly by Newcomen, and above all by James Watt. The expiration of Watt's vital patent occurred in 1800, and he himself then retired from the active supervision of his engineering business, having virtually finished his great life's work on the last year of the century which he had marked for all time by the efforts of his genius.

Similarly we may confidently characterise the locomotive engine as an invention belonging to the first half of the nineteenth century, although tramways on the one hand, and steam-engines on the other hand, were ready for the application of steam transport, and the only work that remained to be accomplished in the half century indicated was the bringing of the two things together. The dynamo, as a factor in human life—or, in other words, the electric current as a form of energy producing power and light—is an invention of the second half of the nineteenth century, although the main principles upon which it was built were worked out prior to the year 1851.

It will be seen, in the course of the subsequent pages, that portable electric power has as yet won its way only into very up-to-date workshops and mines, and that the means by which it will be applied to numerous useful purposes in the field, the road, and the house will be distinctly inventions of the twentieth century. Similarly the steam-engine has not really been placed upon the ordinary road, although efforts have been made for more than a century to put it there, the conception of a road locomotive being, in fact, an earlier one than that of an engine running on rails. Steam automobiles and traction engines are still confined to special purposes, the natures of which prove that certain elements of adaptability are still lacking in order to render them universally useful as are the locomotive and the steam-ship.

In nearly every other important line of human needs and desires it will be found that merely tentative efforts have been made by ingenious minds resulting in inventions of greater or less promise. Many of the finest conceptions which have necessarily been set down as failures have missed fulfilling their intended missions, not so much by reason of inherent weakness, as through the want of accessory circumstances to assist them. As in biology, so in industrial progress the definition of fitness appended to the law of the survival of the fittest must have reference to the environment.

A foolish law or public prejudice results in the temporary failure of a great invention, and the inventor's patent succumbs to the inexorable operation of the struggle for existence. Yet, fortunately for mankind, if not for the individual inventor, an idea does not suffer extinction as the penalty for non-success in the struggle. "The beginning of creation," says Carlyle, "is light," and the kind of light which inventors throw upon the dark problems involving man's industrial progress is providentially indestructible.

Twentieth century inventions—as the term is used in this book—are, therefore, those which are destined to fulfil their missions during the ensuing hundred years. They are those whose light will not only exist in hidden places, but will also shine abroad to help and to bless mankind. Or, if we may revert to the former figure, they are those which have not only been planted in the seed and have germinated in the leaf, but which have grown to goodly proportions, so that none may dare to assert that they have been planted for nought. A man's age is the age in which he does his work rather than that in which he struggles to years of maturity. Moore and Byron were poets of the nineteenth century, although the one had attained to manhood and the other had grown from poverty to inherit a peerage before the new century dawned.

The prophetic rôle—although proverbially an unsafe one—is nevertheless one which every business man must play almost every day of his life. The merchant, the manufacturer, the publisher, the director, the manager, and even the artist, must perforce stake some portion of his success in life upon the chance of his forecast as to the success of a particular speculation, article of manufacture, or artistic conception, and its prospects of proving as attractive or remunerative as he has expected it to be. The successful business man no doubt makes his plans, as far as may be practicable, upon the system indicated by the humorist, who advises people never to prophesy unless they happen to know, but the nature of his knowledge is almost always to some extent removed from certainty. He may spend much time in laborious searching; make many inquiries from persons whom he believes to be competent to advise him; diligently study the conditions upon which the problem before him depends—in short, he may take every reasonable precaution against the chances of failure, yet, in spite of all, he must necessarily incur risks. And so it is with regard to the task of forecasting the trend of industrial improvement. All who are called upon to lay their plans for a number of years beforehand must necessarily be deeply interested in the problems relating to the various directions which the course of that improvement may possibly take. Meanwhile their estimates of the future, although based upon an intimate knowledge of the past and aided by naturally clear powers of insight, must be hypothetical and conditional. Unfortunately for the vast majority of manufacturing experts, the thoroughness with which they have mastered the details of one particular branch of industry too often blinds them to the chances of change arising from localities beyond their own restricted fields of vision.

The merriment occasioned by the first proposals for affixing pneumatic tyres to bicycles may be cited as a striking instance of the lack of forecasting insight displayed by very many of those who are best entitled to pronounce opinions on the minutiae of their particular avocations. In almost every "bike" shop and factory throughout the United Kingdom and America, the suggestion of putting an air-filled hosepipe around each wheel of the machine to act as a tyre was received with shouts of ridicule!

Railway men, who understood the wonderful elasticity imparted by air to pieces of mechanism, such as the pneumatic brake, were not by any means so much inclined to laughter; but naturally, for the most part, they deferred to the rule which enjoins every man to stick to his trade. The rule in question — when applied to the task of estimating the worth of inventions claiming to produce revolutionary effects in any industry — is necessarily, in the majority of cases, more or less irrelevant, because such an invention should be regarded not so much as a proposed *innovation* in an old trade as the *creation* of a new one.

George Stephenson's ideas on the transport of passengers and goods were almost unanimously condemned by the experts of his day who were engaged in that line of business. On points relating to wheels of waggons and the harness of horses, the opinions of these men were probably worth something; but in relation to steam locomotives, carriages and trucks running upon rails, their judgment was not merely worthless, but a good deal worse; it was indeed actually misleading, because based on a pretence of knowledge of a trade which was to be called into existence to compete with their own. "Great is Diana of the Ephesians" said the artificers of old; and on the strength of their expert knowledge in the making of idols they set themselves up as judges of systems of theology and morality. The argument, although based on self-interest subjectively, was nevertheless intended to carry weight even among persons who wished to judge the questions in dispute according to their merits, and most of the latter were only too ready to accept the implied dictum that men who work about a temple must be experts in theology! The principles upon which Royal Commissions and Select Committees are sometimes appointed and entrusted with the onerous duty of deciding upon far-reaching industrial problems, affecting the progress of trade and manufactures in the present day, involve exactly the same kind of fallacy. Men are selected to pronounce judgment upon the proposals of their rivals in trade, and narrow-minded specialists to give their opinions upon projects which essentially belong to the border lands between two or more branches of industry, and cannot be understood by persons not possessing a knowledge of both.

Yet the world's work goes on apace; and as capital is accumulated and seeks to find new outlets the multiplication of industrial projects must continue in spite of every discouragement. This process will go on at a rate even faster than that which was exhibited at the beginning of the nineteenth century; but in watching the course of advancement, the world must take count of ideas rather than of the names of those who may have claims to rank as the originators of ideas. While for purposes of convenience, history labels certain great inventive movements, each with the name of one pre-eminent individual who has contributed largely to its success, nothing like a due appraisement of the services rendered by other men is ever attempted. It is not even as if the commanding general should by public acclamation receive all the applause for a successful campaign to the exclusion of his lieutenants. The pioneers in each great department of invention have generally acted as forerunners of the men whose names have become the most famous. They have borne much of the heat and burden of the day, while their successors have reaped the fruits of triumph. Mr. Herbert Spencer's strong protest against the part assigned by some writers in the mental and industrial evolution of the human race to the influence of great men is certainly fully justified, if the attribute of greatness is to be ascribed only to those whose names figure in current histories. The parts performed by others, whose fate it may have been to have fallen into comparatively unfavourable environments, may have entitled them even more eminently to the acclamation of greatness.

The world in such a matter asks, reasonably enough under the circumstances, Shall we omit to honour any of the great men who have played important parts in an industrial movement, assigning as our motive the difficulty of enumerating so many names? For the encouragement of those to whom the ambition for fame acts as a great stimulus to self-devotion in the interests of human progress, it is unavoidable that some men should be singled out and made heroes, while the much more numerous class of those who have also done great work, but who have not been quite so successful, must pass out of the ken of all, excepting the few who possess an expert knowledge of the various subjects which they have taken in hand.

Still the distortion to which history has been subjected through its biographical mode of treatment must always be reckoned with as a factor of possible error by any one attempting to read the riddle of the past, and it may offer a still more dangerous snare to one who tries to deduce the future course of events from the evidences of the past, and the promises which they hold out. People are naturally prone to take it for granted that the world's progress during the first part of the twentieth century depends upon the future work of those inventors and industrial promoters whose names have become most famous during the latter half of the nineteenth. But this personal treatment of the subject will be found to be in the last degree unsatisfactory, when judged in the light both of past experience and of some of the utterances of those eminent inventors who have tried to forecast the future in their own particular lines of research.

If, therefore, we look at the whole subject from the entirely impersonal point of view, and face the task of forecasting the progress of industry during the twentieth century, in this aspect we shall find that we have entered upon a chapter in the evolution of the human race—dealing, in fact, with a branch of anthropology. We see certain industrial and inventive forces at work, producing certain initial effects, but plainly, as yet, falling immeasurably short of an entire fulfilment of their possibilities; setting to work a multitude of busy brains, planning and arranging, and gradually preparing the minds of the more apathetic portion of humanity for the reception of new ideas and the adoption of improved methods of life and of work. Whither is it all tending? Will the twentieth century bring about as great a change upon the earth—man's habitat—as the nineteenth did? Or have the possibilities of really great and effective industrial revolutions been practically exhausted? The belief impressed upon the Author's mind, by facts and considerations evoked during the collection of materials for this book, is that the march of industrial progress is only just beginning, and that the twentieth century will witness a far greater development than the nineteenth has seen.

The great majority of mankind still require to be released from the drudgery of irksome, physical exertion, which, when power has been cheapened, will be seen to be to a very large extent avoidable. Pleasurable exercise will be substituted for the monotonous, manual labour which, while it continues, generally precludes the possibility of mental improvement. Hygienic science will insist more strenuously than ever upon the great truth that, in order to be really serviceable in promoting the health of mind and body, physical exertion must be in some degree exhilarating, and the bad old practice of "all work and no play," which was based upon the assumption that a boy can get as much good out of chopping wood for an hour as out of a bicycle ride or a game of cricket, will be relegated to the limbo of exploded fallacies.

The race, as a whole, will be athletic in the same sense in which cultured ladies and gentlemen are at present. It will, a century hence, offer a still more striking contrast to the existing state of the Chinese, who bandage their women's feet in order to show that they are high born and never needed to walk or to exert themselves!—the assumption being that no one would ever move a muscle unless under fear of the lash of poverty or of actual hunger. The farther Western civilisation travels from that effete Eastern ideal, the greater will be the hope for human progress in physical, mental and moral well-being.

CHAPTER II
NATURAL POWER

"Nature," remarked James Watt when he set to work inventing his improved steam-engine, "has always a weak side if we can only find it out." Many invaluable secrets have been successfully explored through the discovery of Nature's "weak side" since that momentous era in the industrial history of the world; and the nineteenth century, as Watt clearly foresaw, has been emphatically the age of steam power. In the condenser, the high pressure cylinder and the automatic cut-off, which utilises the expansive power of steam vapour, mankind now possesses the means of taming a monster whose capacities were almost entirely unknown to the ancients, and of bringing it into ready and willing service for the accomplishment of useful work. Vaguely and loosely it is often asserted that the age of steam is now giving place to that of electricity; but these two cannot yet be logically placed in opposition to one another. No method has yet been discovered whereby the heat of a furnace can be directly converted into an electric current. The steam-engine or, as Watt and his predecessors called it, the "fire-engine" is *par excellence* the world's prime motor; and by far the greater proportion of the electrical energy that is generated to-day owes its existence primarily to the steam-engine and to other forms of reciprocating machinery designed to utilise the expansive power of vapours or gases acting in a similar manner to steam.

The industrial revolutions of the coming century will, without doubt, be brought about very largely through the utilisation of Nature's waste energy in the service of mankind. Waterfalls, after being very largely neglected for two or three generations, are now commanding attention as valuable and highly profitable sources of power. This is only to be regarded as forming the small beginning of a movement which, in the coming century, will "acquire strength by going," and which most probably will, in less than a hundred years, have produced changes in the industrial world comparable to those brought about by the invention of the steam-engine.

Lord Kelvin, in the year 1881, briefly, but very significantly, classified the sources of power available to man under the five primary headings of tides, food, fuel, wind, and rain. Food is the generator of animal energy, fuel that of the power obtained from steam and other mechanical expansive engines; rain, as it falls on the hill-tops and descends in long lines of natural force to the sea coasts, furnishes power to the water-wheel; while wind may be utilised to generate mechanical energy through the agency of windmills and other contrivances. The tides as a source of useful power have hardly yet begun to make their influence felt, and indeed the possibility of largely using them is still a matter of doubt. The relative advantages of reclaiming a given area of soil for purposes of cultivation, and of converting the same land into a tidal basin in order to generate power through the inward and outward flow of the sea-water, were contrasted by Lord Kelvin in the statement of a problem as follows: Which is the more valuable—an agricultural area of forty acres or an available source of energy equal to one hundred horse-power? The data for the solution of such a question are obviously not at hand, unless the quality of the land, its relative nearness to the position at which power might be required, and several other factors in its economic application have been supplied. Still, the fact remains that very large quantities of the coastal land and a considerable quantity of expensive work would be needed for the generation, by means of the tides, of any really material quantity of power.

It is strange that, while so much has been written and spoken about the possibility of turning the energy of the tides to account for power in the service of man, comparatively little attention has been paid to the problem of similarly utilising the wave-power, which goes to waste in such inconceivably huge quantities. Where the tidal force elevates and depresses the sea-water on a shore, through a vertical distance of say eight feet, about once in twelve hours, the waves of the ocean will perform the same work during moderate weather once in every twelve or fifteen seconds. It is true that the moon in its attraction of the sea-water produces a vastly greater sum total of effect than the wind does in raising the surface-waves, but reckoning only that part of the ocean energy which might conceivably be made available for service it is safe to calculate that the waves offer between two and three thousand times as much opportunity for the capture of natural power and its application to useful work as the tides could ever present. In no other form is the energy of the wind brought forward in so small a compass or in so concrete a form. A steam-ship of 10,000 tons gross weight which rises and falls ten times per minute through an average height of 3·3 feet is thereby subjected to an influence equal to 22,400 horse-power. In this estimate the unit of the horse-power which has been adopted is Watt's arbitrary standard of "33,000 foot pounds per minute". The work done in raising the vessel referred to is equal to ten horse-power multiplied by the number of pounds in a ton, or, in other words, 22,400 horse-power, as stated.

Wind-power, again, has been to a large extent neglected since the advent of the steam-engine. The mightiest work carried out in any European country in the early part of the present century was that which the Dutch people most efficiently performed in the draining of their reclaimed land by means of scores of windmills erected along their seaboard. Even to the present day there are no examples of the direct employment of the power of the wind which can be placed in comparison with those still to be found on the coasts of Holland. But, unfortunately for the last generation of windmill builders, the intermittent character of the power to which they had to trust completely condemned it when placed in competition with the handy and always convenient steam-engine. The wind bloweth "where it listeth," but only at such times and seasons as it listeth, and its vagaries do not suit an employer whose wages list is mounting up whether he has his men fully occupied or not. The storage of power was the great thing needful to enable the windmill to hold its own. The electrical storage battery, compressed air, and other agencies which will be referred to later on, have now supplied this want of the windmill builder, but in the meantime his trade has been to a large extent destroyed. For its revival there is no doubt that, as Lord Kelvin remarked in the address already quoted, "the little thing wanted to let the thing be done is cheap windmills."

This, however, leads to another part of the problem. The costliness of the best modern patterns of windmill as now so extensively used, particularly in America, is mainly due to the elaborate, and, on the whole, successful attempts at minimising the objection of the intermittent nature of the source of power. To put the matter in another way, it may be said that lightness, and sensitiveness to the slightest breeze, have had to be conjoined with an eminent degree of safety in the severest gale, so that the most complicated self-regulating mechanisms have been rendered absolutely imperative. Once the principle of storage is applied, the whole of the conditions in this respect are revolutionised. There is no need to attempt the construction of wind-motors that shall run lightly in a soft zephyr of only five or six miles an hour, and stability is the main desideratum to be looked to.

The fixed windmill, which requires no swivel mechanism and no vane to keep it up to the wind, is the cheapest and may be made the most substantial of all the forms of wind-motor. In its rudimentary shape this very elementary windmill resembles a four-bladed screw steam-ship propeller. The wheel may be constructed by simply erecting a high windlass with arms bolted to the barrel at each end, making the shape of a rectangular cross. But those at one end are fixed in such positions that when viewed from the side they bisect the angles made by those at the other side. Sails of canvas or galvanised iron are then fastened to the arms, the position of which is such that the necessary obliquity to the line of the barrel is secured at once.

Looking at this elementary and at one time very popular form of windmill, and asking ourselves what adaptation its general principle is susceptible of in order that it may be usefully employed in conjunction with a storage battery, we find, at the outset, that, inasmuch as the electric generator requires a high speed, there is every inducement to greatly lengthen the barrel and at the same time to make the arms of the sails shorter, because short sails give in the windmill the high rate of speed required.

We are confronted, in fact, with the same kind of problem which met the constructors of turbine steam-engines designed for electric lighting. The object was to get an initial speed which would be so great as to admit of the coupling of the dynamo to the revolving shaft of the turbine steam-motor, without the employment of too much reducing gear. In the case of the wind-motor the eighteenth century miller was compelled to make the arms of his mill of gigantic length, so that, while the centre of the wind pressure on each arm was travelling at somewhere near to the rate of the wind, the axis would not be running too fast and the mill stones would never be grinding so rapidly as to "set the *tems*—or the lighter parts of the corn—on fire."

The dynamo for the generation of the electric current demands exactly the opposite class of conditions. We may therefore surmise that the windmill of the future, as constructed for the purposes of storing power, will have a long barrel upon which will be set numerous very short blades or sails. Reducing this again to its most convenient form, it is plain that a spiral of sheet-metal wound round the barrel will offer the most convenient type of structure for stability and cheapness combined. At the end of this long barrel will be fixed the dynamo, the armature of which is virtually a part of the barrel itself, while the magnets are placed in convenient positions on the supporting uprights. From the generating dynamo the current is conveyed directly to the storage batteries, and these alone work the electric motor, which, if desired, keeps continually in motion, pumping, grinding, or driving any suitable class of machinery.

It is rather surprising to find how relatively small is the advantage possessed by the vane-windmill over the fixed type in the matter of continuity of working. During about two years the Author conducted a series of experiments with the object of determining this point, the fixed windmill being applied to work which rendered it a matter of indifference in which way the wheel ran. With the prevailing winds from the west it ran in one direction, and with those of next degree of frequency, namely from the east, it turned in the reverse direction. The mill, however, was effective although the breeze might veer several points from either of the locations mentioned. It was found that there were rather less than one-fourth of the points of the compass, the winds from which would bring the wheel to a standstill or cause it to swing ineffectively, but as these were the directions in which the wind least frequently blew it might safely be reckoned that not one-eighth of the possible working hours of a swivel-windmill were really lost in the fixed machine.

With the type adapted to the working of a dynamo as already described, it will, in most cases, be convenient to construct two spirals on uprights set in three holes in the ground, forming lines at right angles to each other, but both engaging, by suitable gearing, with the electric current generator situated at the angle. This will be found cheaper than to go to the expense of constructing the mill on a swivel so that it may follow the direction of the wind. At the same time it should be noticed that the adoption of the high speed wind-wheel, consisting of some kind of spiral on a very long axis, may be made effective for improving even the swivel windmill itself, so as to adapt it for electric generation and conservation of power through the medium of the storage battery. Supposing that a number of small oblique sails be set upon an axis lying in the direction of the wind, the popular conception of the result of such an arrangement is that the foremost sails would render those behind it almost, if not entirely, useless.

The analogy followed in reaching this conclusion is that of the sails of a ship, but, as applied to wind-motors, it is quite misleading, because not more than one-third or one-fourth of the energy of the wind is expended upon the oblique sails of an ordinary wind-wheel. Moreover, in the case of a number of such wheels set on a long axis, one behind the other as described, the space within which the shelter of the front sail is operative to keep the wind from driving the next one is exceedingly minute.

The elasticity of the air and its frictional inertia when running in the form of wind cause the current to proceed on its course after a very slight check, which in point of time is momentary and in its effects almost infinitesimal. This being the case, and the principal expense attendant upon the construction of ordinary wind-engines being due to the need for providing a large diameter of wind-wheel, with all the attendant complications required to secure such a wheel from risk, it is obvious that as soon as the long axis and the very short sail, or the metallic spiral, have been generally introduced as adjuncts to the dynamo storage battery, an era of cheaper wind-motors will have been entered upon,—in fact, the "little want" of which Lord Kelvin spoke in 1881 will have been supplied. The high speed which the dynamo requires, and the more rapid rate at which windmills constructed on this very economical principle must necessarily run, both mark the two classes of apparatus as being eminently suited for mutual assistance in future usefulness.

The anemometer of the "Robinson" type, having four little hemispherical cups revolving horizontally, furnishes the first hint of another principle of construction adapted to the generation of electricity. Some years ago a professor in one of the Scottish Universities set up a windmill which was simply an amplified anemometer, and connected it with several of Faure's storage batteries for the purpose of furnishing the electric light to his residence. His report regarding his experience with this arrangement showed that the results of the system were quite satisfactory.

In this particular type of natural motor the wind-wheel, of course, is permanently set to run no matter from what direction the wind may be blowing. Tests instituted with the object of determining the pressure which the wind exerts on the cup of a "Robinson" anemometer have shown that when the breeze blows into the concave side of the cup, its effect is rather more than three times as strong as when it blows against the convex side. At any given time the principal part of the work done by a windmill constructed on this principle is being carried out by one cup which has its concave side presented to the wind, while, opposite to it, there is another cup travelling in the opposite direction to that of the wind but having its convex side opposed.

The facts that practically only one sail of the mill is operative at any given time, and that even the work which is done by this must be diminished by nearly one-third owing to the opposing "pull" of the cup at the opposite side, no doubt must detract from the merits of such a wind-motor, judged simply on the basis of actual area of sail employed. But when the matter of cost alone is taken as the standard, the advantages are much more evenly balanced than they might at first sight seem to be.

The cup-shaped sail may be greatly improved upon for power-generating purposes by adopting a sail having a section not semicircular but triangular in shape, and by extending its length in the vertical direction to a very considerable extent. Practically this cheap and efficient wind-motor then becomes a square or hexagonal upright axis of fairly large section, to each side of which is secured a board or a rigid sheet-metal sail projecting beyond the corners. The side of the axis and the projecting portion of the sail then together form the triangular section required.

For the sake of safety in time of storm, an opening may be left at the apex of the angle which is closed by a door kept shut through the tension of a spring. When the wind rises to such a speed as to overbalance the force of the spring each door opens and lets the blast pass through. One collateral advantage of this type of windmill is that it may be made to act virtually as its own stand, the only necessity in its erection being that it should have a collar fitting round the topmost bearing, which collar is fastened by four strong steel ropes to stakes securely set in the ground. The dynamo is then placed at the lower bearing and protected from the weather by a metal shield through which the shaft of the axis passes.

For pumping, and for other simple purposes apart from the use of the dynamo, a ready application of this form of wind-engine with a minimum of intricacy or expense may be worked out by setting the lower bearing in a round tank of water kept in circular motion by a set of small paddles working horizontally. Into the water a vertically-working paddle-wheel dips, carrying on its shaft a crank which directly drives the pump. This simple wind-motor is particularly safe in a storm, because on attaining a high speed it merely "smashes" the water in the tank.

Solar heat is one of the principal sources of the energy to be derived from the wind. Several very determined and ingenious attempts at the utilisation of the heat of sunshine for the driving of a motor have been made during the past century. As a solution of a mechanical and physical puzzle, the arrangement of a large reflector, with a small steam-boiler at the focus of the heat rays thrown by it, is full of interest. Yet, when a man like the late John Ericsson, who did so much to improve the caloric engine, and the steam-ship as applied to war-like purposes, meets with failure in the attempt to carry such an idea to a commercially successful issue, there is at least *prima facie* evidence of some obstacle which places the proposed machine at a disadvantage in competition with its rivals.

The solar engine, if generally introduced, would be found more intermittent in its action than the windmill—excepting perhaps in a very few localities where there is a cloudless sky throughout the year. The windmill gathers up the power generated by the expansion of the air in passing over long stretches of heated ground, while a solar engine cannot command more of the sun's heat than that which falls upon the reflector or condenser of the engine itself. The latter machine may possibly have a place assigned to it in the industrial economy of the future, but the sum total of the power which it will furnish must always be an insignificant fraction.

The wave-power machine, when allied to electric transmission, will, without doubt, supply in a cheap and convenient form a material proportion of the energy required during the twentieth century for industrial purposes. Easy and effective transmission is a *sine quâ non* in this case, just as it is in the utilisation of waterfalls situated far from the busy mart and factory. Hardly any natural source of power presents so near an approach to constancy as the ocean billows. Shakespeare takes as his emblem of perpetual motion the dancing "waves o' th' sea".

But the ocean coasts—where alone natural wave-power is constant—are exactly the localities at which, as a rule, it is the least practicable to build up a manufacturing trade. Commerce needs smooth water for the havens offered to its ships, and inasmuch as this requirement is vastly more imperative during the early stages of civilisation than cheap power, the drift of manufacturing centres has been all towards the calm harbours and away from the ocean coasts. But electrical transmission in this connection abolishes space, and can bring to the service of man the power of the thundering wave just as it can that of the roaring torrent or waterfall.

The simplest form of wave-motor may be suggested by the force exerted by a ferry boat or dinghy tied up to a pier. The pull exerted by the rope is equal to the inertia of the boat as it falls into the trough of each wave successively, and the amount of strain involved in rough weather may be estimated from the thickness of the rope that is generally found necessary for the security of even very small craft indeed. A similar suggestion is conveyed by the need for elaborate "fenders" to break the force of the shock when a barge is lying alongside of a steamer, or when any other vessel is ranging along a pier or jetty.

A buoy of large size, moored in position at a convenient distance from a rock-bound ocean coast, will supply the first idea of a wave-motor on this primary principle as adapted for the generation of power. On the cliff a high derrick is erected. Over a pulley or wheel on the top of this there is passed a wire-rope cable fastened on the seaward side to the buoy, and on the landward side to the machinery in the engine-house. The whole arrangement in fact is very similar in appearance to the "poppet-head" and surface buildings that may be seen at any well-equipped mine. The difference in principle, of course, is that while on a mine the engine-house is supplying power to the other side of the derrick, the relations are reversed in the wave-motor, the energy being passed from the sea across into the engine-house. The reciprocating, or backward and forward, movement imparted to the cable by the rising and falling of the buoy now requires to be converted into a force exerted in one direction. In the steam-engine and in other machines of similar type, the problem is simplified by the uniform length of the stroke made by the piston, so that devices such as the crank and eccentric circular discs are readily applicable to the securing of a rotatory motion for a fly-wheel from a reciprocating motion in the cylinders. In the application of wave-power provision must be made for the utilisation of the force derived from movements of *differing lengths*, as well as of *differing characters*, in the force of impact. Every movement of the buoy which imparts motion to the pulley on top of the derrick must be converted into an additional impetus to a fly-wheel always running in the same direction.

The spur-wheel and ratchet, as at present largely used in machinery, offer a rough and ready means of solving this problem, but two very important improvements must be effected before full advantage can be taken of the principle involved. In the first place it is obvious that if a ratchet runs freely in one direction and only catches on the tooth of the spur-wheel when it is drawn in the other, the power developed and used is concentrated on one stroke, when it might, with greater advantage, be divided between the two; and in the second place the shock occasioned by the striking of the ratchet against the tooth when it just misses catching one of the teeth and is then forced along the whole length of the tooth gathering energy as it goes, must add greatly to the wear and tear of the machinery and to the unevenness of the running.

Taking the first of these difficulties into consideration it is obvious that by means of a counterbalancing weight, about equal to half that of the buoy, it is possible to cause the wave-power to operate two ratchets, one doing work when the pull is to landwards and the other when it is to seawards. Each, however, must be set to catch the teeth of its own separate spur-wheel; and, inasmuch as the direction of the motion in one case is different from what it is in the other, it is necessary that, by means of an intervening toothed wheel, the motion of one of these should be reversed before it is communicated to the fly-wheel. The latter is thus driven always in the same direction, both by the inward and by the outward stroke or pull of the cable from the buoy.

Perhaps the most convenient development of the system is that in which the spur-wheel is driven by two vertically pendant toothed bands, resembling saws, and of sufficient length to provide for the greatest possible amplitude of movement that could be imparted to them by the motion of the buoy. The teeth are set to engage in those of the spur-wheel, one band on each side, so that the effective stroke in one case is downward, while in the other it is upward. These toothed bands are drawn together at their lower ends by a spring, and they are also kept under downward tension by weights or a powerful spring beneath. The effect of this is that when both are drawn up and down the spur-wheel goes round with a continuous motion, because at every stroke the teeth of one band engage in the wheel and control it, while those of the reversed one (at the other side) slip quite freely.

The shock occasioned by the blow of the ratchet on the spur-wheel, or of one tooth upon another, may be reduced almost to vanishing point by multiplying the number of ratchets or toothed bands, and placing the effective ends, which engage in the teeth of the wheel successively, one very slightly in advance of the other. In this way the machine is so arranged that, no matter at what point the stroke imparted by the movement of the buoy may be arrested, there is always one or other of the ratchets or of the teeth which will fall into engagement with the tooth of the spur-wheel, very close to its effective face, and thus the momentum acquired by the one part before it impinges upon the other becomes comparatively small.

The limit to which it may be practicable to multiply ratchets or toothed bands will, of course, depend upon the thickness of the spur-wheel, and when this latter has been greatly enlarged, with the object of providing for this feature, it becomes virtually a steel drum having bevelled steps accurately cut longitudinally upon its periphery.

The masts of a ship tend to assume a position at right angles to the water-line. When the waves catch the vessel on the beam the greatest degree of pendulous swing is brought about in a series of waves so timed, and of such a length, that the duration of the swing coincides with the period required for one wave to succeed another. The increasing slope of the ship's decks, due to the inertia of this continuous rhythmical motion, often amounts to far more than the angle made by the declivity of the wave as compared with the sea level; and it is, of course, a source of serious danger in the eyes of the mariner.

But, for the purposes of the mechanician who desires to secure power from the waves, the problem is not how to avoid a pendulous motion but how to increase it. For each locality in which any large wave-power plant of machinery is to be installed, it will therefore be advisable to study the characteristic length of the wave, which, as observation has proved, is shorter in confined seas than in those fully open to the ocean. It is advisable then to make the beam width of the buoy, no matter how it may be turned, of such a length that when one side is well in the trough of a wave the other must be not far from the crest.

Practically the best design for such a floating power-generator will be one in which four buoys are placed, each of them at the end of one arm of a cross which has been braced up very firmly. From the angle of intersection projects a vertical mast, also firmly held by stays or guys. The whole must be anchored to the bottom of the sea by attachment to a large cemented block or other heavy weight having a ring let into it, from which is attached a chain of a few links connecting with an upright beam. It is the continuation of the latter above sea-level which forms the mast. On this beam the framework of the buoy must be free to move up and down.

At first sight it might seem as if this arrangement rendered nugatory the attempt to take advantage of the rise and fall of the buoy; but it is not so when the relations of the four buoys to one another are considered. Although the frame is free to move up and down upon the uprising shaft, still its inclination to the vertical is determined by the direction of the line drawn from a buoy in the trough of a wave to one on the crest. In order to facilitate the free movement, and to render the rocking effect more accurate and free from vibration, sets of wheels running on rails fixed to the beam are of considerable advantage.

The rise and fall of the tides render necessary the adoption of some such compensating device as that which has been indicated. Of course it would be possible to provide for utilising the force generated by a buoy simply moored direct to a ring at the bottom by means of a common chain cable; but this latter would require to be of a length sufficient to provide for the highest possible wave on the top of the highest tide. Then, again, the loose chain at low tide would permit the buoy to drift abroad within a very considerable area of sea surface, and in order to take advantage of the rise and fall on each wave it would be essential to provide at the derrick on the shore end of the wave-power plant very long toothed bands or equivalent devices on a similarly enlarged scale.

By providing three or four chains and moorings, meeting in a centre at the buoy itself but fastened to rings secured to weights at the bottom at a considerable distance apart, the lateral movement might, no doubt, be minimised; and for very simple installations this plan, associated with the device of taking a cable from the buoy and turning it several times round a drum on shore, could be used to furnish a convenient source of cheap power. The drum may carry a crank and shaft, which works the spur-wheel and toothed bands as already described, so that no matter at what stage in the revolution of the drum an upward or downward stroke may be stopped, the motion will still be communicated in a continuous rotary form to the fly-wheel.

But the beam and sliding frame, with buoys, give the best practical results, especially for large installations. It is in some instances advisable, especially where the depth of the water at a convenient distance from the shore is very considerable, not to provide a single beam reaching the whole distance to the bottom, but to anchor an air-tight tank below the surface and well beneath the depth at which wave disturbance is ever felt. From this submerged tank, which approximately keeps a steady position in all tides and weathers, the upward beam is attached by a ring just as would be done if the tank itself constituted the bottom.

One main reason for this arrangement is that the resistance of the beam to the water as it rocks backwards and forwards wastes to some extent the power generated by the force of the waves; and the greater the length of the beam, the longer must be the distance through which it has to travel when the buoys draw it into positions vertical to that of the framework. A thin steel pipe offers less resistance than a wooden beam of equal strength, besides facilitating the use of a simple device for enabling the frame and buoys to slide easily up and down.

The generally fatal defect of those inventions which have been designed in the past with the object of utilising wave-power has arisen from the mistake of placing too much of the machinery in the sea. The device of erecting in the water an adjustable reservoir to catch the wave crests and to use the power derived from them as the water escaped through a water-wheel was patented in 1869. Nearly twenty years later another scheme was brought out depending upon the working of a large pump fixed far under the surface, and connected with the shore so that, when operated by the rising and falling of floats upon the waves, it would drive a supply of water into an elevated reservoir on shore, from which, on escaping down the cliff, the pressure of the water would be utilised to work a turbine.

Earlier devices included the building of a mill upon a rocking barge, having weights and pulleys adjusted to run the machinery on board; and also a revolving float so constructed that each successive wave would turn one portion, but the latter would then be held firm by a toothed wheel and ratchet until another impulse would be given to it in the same direction. This plan included certain elements of the simple system already described; but it is obvious that some of its floating parts might with advantage have been removed to the shore end, where they would not only be available for ready inspection and adjustment, but also be out of harm's way in rough weather.

Different wave-lengths, as already explained, correspond to various periods in the pendulous swing of floating bodies. Examples have been cited by Mr. Vaughan Cornish, M. Sc., in *Knowledge*, 2nd March, 1896, as follows: "A wave-length of fifty feet corresponds to a period of two and a half seconds, while one of 310 feet corresponds to five and a half seconds. It is mentioned that the swing of the steam-ship *Great Eastern* took six seconds." Other authorities state that during a storm in the Atlantic the velocity of the wave was determined to be thirty-two miles an hour, and that nine or ten waves were included in each mile; thus about five would pass in each minute. But in average weather the number of waves to the mile is considerably larger, say, from fifteen to twenty to the mile; and in nearly calm days about double those numbers.

One interesting fact, which gives to wave-power a peculiarly enhanced value as a source of stored wind-power, is that the surface of the ocean—wild as it may at times appear—is not moved by such extremes of agitation as the atmosphere. In a calm it is never so inertly still, and in a storm it is never so far beyond the normal condition in its agitation as is the wind. The ocean surface to some extent operates as the governor of a steam-engine, checking an excess in either direction. In very moderate weather the number of waves to the mile is greatly increased, while their speed is not very much diminished. Indeed the rate at which they travel may even be increased.

This latter phenomenon generally occurs when long ocean rollers pass out of a region of high wind into one of relative calm, the energy remaining for a long time comparatively constant by reason of the multiplication of short, low waves created out of long, high ones. On all ocean coasts the normal condition of the surface is governed by this law, and it follows that, no matter what the local weather may be at any given time, there is always plenty of power available.

An attempt was made by M. C. Antoine, after a long series of observations, to establish a general relation between the speed of the wind and that of the waves caused by it, the formulæ being published in the *Revue Nautique et Coloniale* in 1879. The rule may be taken as correct within certain limits, although in calm weather, when the condition of the ocean surface is almost entirely ruled by distant disturbances, it has but little relevancy. Approximately, the velocity of wave transmission is seven times the fourth root of the wind-speed; so that when the latter is a brisk breeze of sixteen miles an hour the waves will be travelling fourteen miles an hour, or very nearly as fast as the wind. When, on the other hand, a light breeze of nine miles an hour is driving the waves, the latter, according to the formula, should run about twelve and a half miles an hour; but, in point of fact, the influence of more distant commotion nearly always interferes with this result.

As a matter of experience, the waves on an ocean coast are usually running faster than the wind, and, being so much more numerous in calm than they are in rough weather, they maintain comparatively a uniform sum total of energy. It is obvious that, so far as practical purposes are concerned, three waves of an available height of three feet each are as effective as one of nine feet. If the state of the weather be such that the average wave length is 176 feet there will be exactly thirty waves to the mile, and if the speed be twelve miles an hour—that is to say, if an expanse of twelve miles of waves pass a given point hourly—then 360 waves will pass every sixty minutes, or six every minute. In the wave-power plant as described, each buoy of one hundred tons displacement when raised and depressed, say, three feet by every wave will thus be capable of giving power equal to three times 600, or 1,800 foot-tons per minute.

The unit of nominal horse-power being 33,000 foot-pounds or about fifteen foot-tons per minute, it is evident that each buoy, at its maximum, would be capable of giving about 120 horse-power. Supposing that half of the possible energy were exerted at the forward and half at the backward stroke and that each buoy were always in position to exert its full power upon the uprising shaft without deduction, the total effective duty of a machine such as has been described would be 480 horse-power. In practice, however, the available duty would probably, according to minor circumstances, be rather more or rather less than 300 horse-power.

CHAPTER III
STORAGE OF POWER

The three principal forms of stored power which are now in sight above the horizon of the industrial outlook are the electric storage battery, compressed air, and calcium-carbide. The first of these has come largely into use owing to the demand for a regulated and stored supply of electricity available for lighting purposes. Indeed the storage battery has practically rendered safe the wide introduction of electric lighting, because a number of cells, when once charged, are always available as a reserve in case of any failure in the power or in the generators at any central station; and also because, by means of the storage cells or "accumulators," the amount of available electrical energy can be subdivided into different and subordinate circuits, thus obviating the necessity for the employment of currents of very high voltage and eluding the only imperfectly-solved problem of dividing a current traversing a wire as conveniently as lighting gas is divided by taking small pipes off from the gas mains.

Compressed air for the storage of power has hitherto been best appreciated in mining operations, one of the main reasons for this being that the liberated air itself—apart from the power which it conveyed and stored—has been so great a boon to the miner working in ill-ventilated stopes and drives. The cooling effects of the expansion, after close compression, are also very grateful to men labouring hard at very great depths, where the heat from the country rock would become, in the absence of such artificial refrigeration, almost overpowering. For underground railway traffic exactly the same recommendations have, at one period during the fourth quarter of the nineteenth century, given an adventitious stimulus to the use of compressed air.

Yet it is now undoubted that, even in deep mining, the engineer's best policy is to adopt different methods for the conveyance and storage of power on the one hand, and for the ventilation of the workings on the other. Few temptations are more illusory in the course of industrial progress than those presented by that class of inventions which aim at "killing two birds with one stone". If one object be successfully accomplished it almost invariably happens that the other is indifferently carried out; but the most frequent result is that both of them suffer in the attempt to adapt machinery to irreconcilable purposes.

The electric rock-drill is now winning its way into the mines which are ventilated with comparative ease as well as into those which are more difficult to supply with air. It is plain, therefore, that on its merits as a conveyer and storer of power the electric current is preferable to compressed air. The heat that is generated and then dissipated in the compression of any gas for such a purpose represents a very serious loss of power; and it is altogether an insufficient excuse to point to the compensation of coolness being secured from the expansion. Fans driven by electric motors already offer a better solution of the ventilation difficulty, and the advantages on this side are certain to increase rather than to diminish during the next few years.

The electric rock-drill, which can already hold its own with that driven by compressed air, is therefore bound to gain ground in the future. This is a type and indication of what will happen all along the industrial line, the electric current taking the place of the majority of other means adopted for the transmission of power. Even in workshops—where it is important to have a wide distribution of power and each man must be able to turn on a supply of it to his bench at any moment—shafting is being displaced by electric cables for the conveyance of power to numerous small motors.

The loss of power in this system has already been reduced to less than that which occurs with shafting, unless under the most favourable circumstances; and in places where the works are necessarily distributed over a considerable area the advantage is so pronounced that hardly any factories of that kind will be erected ten years hence without resort being had to electricity, and small motors as the means of distributing the requisite supplies of power to the spots where they are needed. It was a significant fact that at the Paris Exposition of 1900 the electric system of distribution was adopted.

In regard to compressed air, however, it seems practically certain that, notwithstanding its inferiority to electric storage of power, it is applicable to so many kinds of small and cheap installations that, on the whole, its area of usefulness, instead of being restricted, will be largely increased in the near future. There will be an advance all along the line; and although electric storage will far outstrip compressed air for the purposes of the large manufacturer, the air reservoir will prove highly useful in isolated situations, and particularly for agricultural work.

For example, as an adjunct to the ordinary rural windmill for pumping water, it will prove much more handy and effective than the system at present in vogue of keeping large tanks on hand for the purpose of ensuring a supply of water during periods of calm weather. Regarding a tank of water elevated above the ground and filled from a well as representing so much stored energy, and also comparing this with an equal bulk of air compressed to about 300 pounds pressure to the square inch, it would be easy to show that—unless the water has been pumped from a very deep well—the power which its elevation indicates must be only a small fraction of that enclosed in the air reservoir.

It will be one great point in favour of compressed air, as a form of stored energy for the special purpose of pumping, that by making a continuous small flow of air take the place of the water at the lowest level in the upward pipe, it is possible to cause it to do the pumping without the intervention of any motor.

One means of effecting this may be simply indicated. The air under pressure is admitted from a very small air pipe and the bubbles, as they rise, fill the hollow of an inverted iron cup rising and falling on a bearing like a hinge. Above and beneath the chamber containing this cup are valves opening upwards and similar to those of an ordinary force or suction pump. The cup must be weighted with adjustable weights so that it will not rise until quite full of air. When that point is reached the stroke is completed, the air having driven upwards a quantity of water of equal bulk with itself, and, as the cup falls again by its own weight, the vacuum caused by the air escaping upwards through the pipe is filled by an inrush of water through the lower valve. The function of the upper valve, at that time, is to keep the water in the pipe from falling when the pressure on the column is removed. The expansive power of the air enables it to do more lifting at the upper than at the lower level, so that a larger diameter of pipe can be used at the former place.

Cheap motors working on the same principle—that is to say through the upward escape of compressed air, gas or vapour filling a cup and operating it by its buoyancy, or turning a wheel in a similar manner—will doubtless be a feature in the machine work of the future; and for motors of this description it is obvious that compressed air will be very useful as the form of power-storage. Excepting under very special conditions, steam is not available for such a purpose, seeing that it condenses long before it has risen any material distance in a column of cold water.

"The present accumulator," remarked Prof. Sylvanus P. Thompson in the year 1881, referring to the Faure storage batteries then in use, "probably bears as much resemblance to the future accumulator as a glass bell-jar used in chemical experiments for holding gas does to the gasometer of a city gasworks, or James Watt's first model steam-engine does to the engines of an Atlantic steamer." When Faure, having in 1880 improved upon the storage battery of Planté, sent his four-cell battery from Paris to Glasgow, carrying in it stored electrical energy, it was found to contain power equal to close upon a million foot-pounds, which is about the work done by a horse-power during the space of half an hour. This battery weighed very nearly 75 lb. It nevertheless represented an immense forward step in the problem of compressing a given quantity of potential power into a small weight of accumulator.

The progress made during less than twenty years to the end of the century may be estimated from the conditions laid down by the Automobile Club of Paris for the competitive test of accumulators applicable to auto-car purposes in 1899. It was stipulated that five cells, weighing in all 244 lb., should give out 120 ampere-hours of electric intensity; and that at the conclusion of the test there should remain a voltage of 1·7 volt per cell.

Very great improvements in the construction of electric accumulators are to be looked for in the near future. Hitherto the average duration of the life of a storage cell has not been more than about two years; and where impurities have been present in the sulphuric acid, or in the litharge or "minium" employed, the term of durability has been still further shortened. It must be remembered that while the principal chemical and electrical action in the cell is a circular one,—that is to say, the plates and liquids get back to the original condition from which they started when beginning work in a given period,—there is also a progressive minor action depending upon the impurities that may be present. Such a reagent, for instance, as nitric acid has an extremely injurious effect upon the plates.

During the first decade after Planté and Faure had made their original discoveries, the main drawback to the advancement of the electric accumulator for the storage of power owed its existence to the lack of precise knowledge, among those placed in charge of storage batteries, as to the destructive effects of impurities in the cells. It is, however, now the rule that all acids and all samples of water used for the purpose must be carefully tested before adoption, and this practice, in itself, has greatly prolonged the average life of the accumulator cell.

The era of the large electric accumulator of the kind foreshadowed by Prof. Sylvanus P. Thompson has not yet arrived, the simple reason being that electric power storage—apart from the special purposes of the subdivision and transmission for lighting—has not yet been tried on a large scale. For the regulation and graduation of power it is exceedingly handy to be able to "switch-on" a number of small accumulator cells for any particular purpose;

and, of course, the degree of control held in the hands of the engineer must depend largely on the smallness of each individual cell, and the number which he has at command. This fact of itself tends to keep down the size of the storage cell which is most popular.

But when power storage by means of the electric accumulator really begins in earnest the cells will attain to what would at present be regarded as mammoth proportions; and the special purpose aimed at in each instance of power installation will be the securing of continuity in the working of a machine depending upon some intermittent natural force. Windmills are especially marked out as the engines which will be used to put electrical energy into the accumulators. From these latter again the power will be given out and conveyed to a distance continuously.

High ridges and eminences of all kinds will in the future be selected as the sites of wind-power and accumulator plants. In the eighteenth century, when the corn from the wheat-field required to be ground into flour by the agency of wind-power, it was customary to build the mill on the top of some high hill and to cart all the material laboriously to the eminence. In the installations of the future the power will be brought to the material rather than the material to the power. From the ranges or mountain peaks, and also from smaller hills, will radiate electrical power-nerves branching out into network on the plains and supplying power for almost every purpose to which man applies physical force or electro-chemical energy.

The gas-engine during the twentieth century will vigorously dispute the field against electrical storage; and its success in the struggle—so far as regards its own particular province—will be enhanced owing to the fact that, in some respects, it will be able to command the services of electricity as its handmaid. Gas-engines are already very largely used as the actuators of electric lighting machinery. But in the developments which are now foreshadowed by the advent of acetylene gas the relation will be reversed. In other words, the gas-engine will owe its supply of cheap fuel to the electric current derived at small expense from natural sources of power.

Calcium carbide, by means of which acetylene gas is obtained as a product from water, becomes in this view stored power. The marvellously cheap "water-gas" which is made through a jet of steam impinging upon incandescent carbons or upon other suitable glowing hot materials will, no doubt, for a long time command the market after the date at which coal-gas for the generation of power has been partially superseded.

But it seems exceedingly probable that a compromise will ultimately be effected between the methods adopted for making water-gas and calcium carbide respectively, the electric current being employed to keep the carbons incandescent. When power is to be sold in concrete form it will be made up as calcium carbide, so that it can be conveyed to any place where it is required without the assistance of either pipes or wires. But when the laying of the latter is practicable—as it will be in the majority of instances—the gas for an engine will be obtainable without the need for forcing lime to combine with carbon as in calcium carbide.

Petroleum oil is estimated to supply power at just one-third the price of acetylene gas made with calcium carbide at a price of £20 per ton. This calculation was drawn up before the occurrence of the material rise in the price of "petrol" in the last year of the nineteenth century; while, concurrently, the price of calcium carbide was falling. A similar process will, on the average, be maintained throughout each decade; and, as larger plants, with cheaper natural sources of energy, are brought into requisition, the costs of power, as obtained from oil and from acetylene gas, will more and more closely approximate, until, in course of time, they will be about equal; after which, no doubt, the relative positions will be reversed, although not perhaps in the same ratio. Time is all on the side of the agent which depends for its cheapness of production on the utilisation of any natural source of power which is free of all cost save interest, wear and tear, and supervision.

Even the steam-engine itself is not exempt from the operation of the general law placing the growing advantage on the side of power that is obtainable gratis. One cubic inch of water converted into steam and at boiling point will raise a ton weight to the height of one foot; and the quantity of coal of good quality needed for the transformation of the water is very small. One pound of good coal will evaporate nine pounds of water, equal to about 250 cubic inches, this doing 250 foot-tons of work. But Niagara performs the same amount of work at infinitely less cost. However small any quantity may be, its ratio to nothing is infinity.

It has been the custom during the nineteenth century to institute comparisons between the marvellous economy of steam power and the expensive wastefulness of human muscular effort. For instance, the full day's work of an Eastern porter, specially trained to carry heavy weights, will generally amount to the removal of a load of from three to five hundred-weight for a distance of one mile; but such a labourer in the course of a long day has only expended as much power as would be stored up in about five ounces of coal.

Still the fact remains that one of the greatest problems of the future is that which concerns the reduction in the cost of power. Hundreds of millions of the human race pass lives of a kind of dull monotonous toil which develops only the muscular, at the expense of the higher, faculties of the body; they are almost entirely cut off from social intercourse with their fellow-men, and they sink prematurely into decrepitude simply by reason of the lack of a cheap and abundant supply of mechanical power, ready at hand wherever it is wanted. Scores of "enterprises of great pith and moment" in the industrial advancement of the world have to be abandoned by reason of the same lack. In mining, in agriculture, in transport and in manufacture the thing that is needful to convert the "human machine" into a more or less intelligent brainworker is cheaper power. All the technical education in the world will not avail to raise the labourer in the intellectual scale if his daily work be only such as a horse or an engine might perform.

The transmission of power through the medium of the electric current will naturally attain its first great development in the neighbourhoods of large waterfalls such as Niagara. When the manufacturers within a short radius of the source of power in each case have begun to fully reap the benefit due to cheap power, competition will assert itself in many different ways. The values of real property will rise, and population will tend to become congested within the localities' served.

It will be found, however, that facilities for shipment will to a large extent perpetuate the advantage at present held by manufactories situated on ports and harbours; and this, of course, will apply with peculiar force to the cases of articles of considerable bulk. Where a very great deal of power is needed for the making of an article or material of comparatively small weight and bulk proportioned to its value—such for instance as calcium carbide or aluminium—the immediate vicinity of the source of natural power will offer superlative inducements. But an immense number of things lie between the domains of these two classes, and for the economical manufacture of these it is imperative that both cheap power and low wharfage rates should be obtainable.

An increasingly intense demand must thus spring up for systems of long distance transmission, and very high voltage will be adopted as the means of diminishing the loss of power due to leakage from the cables. Similarly the "polyphase" system—which is eminently adapted to installations of the nature indicated—must demand increasing attention.

Taking a concrete example, mention may be made of the effects to be expected from the proposed scheme for diverting some of the headwaters of the Tay and its lakes from the eastern to the western shores of Scotland and establishing at Loch Leven—the western inlet, not the inland lake of that name—a seaport town devoted to manufacturing purposes requiring very cheap supplies of power. It is obvious that the owners of mills in and around Glasgow, and only forty or fifty miles distant, will make the most strenuous exertions to enable them to secure a similar advantage.

It is already claimed that with the use of currents of high voltage for carrying the power, and "step-down transformers" converting these into a suitable medium for the driving of machinery, a fairly economical transmission can be ensured along a distance of 100 miles. It therefore seems plain that the natural forces derived from such sources as waterfalls can safely be reckoned upon as friends rather than as foes of the vested interests of all the great cities of the United Kingdom.

The possibilities of long distance transmission are greatly enhanced by the very recent discovery that a cable carrying a current of high voltage can be most effectually insulated by encasing it in the midst of a tube filled with wet sawdust and kept at a low temperature, preferably at the freezing point of water.

Wireless transmission of a small amount of power has been proved to be experimentally possible. In the rarefied atmosphere at a height of five or ten miles from the earth's surface, electric discharges of very high voltage are conveyed without any other conducting medium than that of the air. By sending up balloons, carrying suspended wires, the positions of despatch and of receipt can be so elevated that the resistance of the atmosphere can be almost indefinitely diminished. In this way small motors have been worked by discharges generated at considerable distances, and absolutely without the existence of any connection by metallic conductors. Possibilities of the exportation of power from suitable stations—such as the neighbourhoods of waterfalls—and its transmission for distances of hundreds or even thousands of miles have been spoken of in relation to the industrial prospects of the twentieth century.

Comparing any such hypothetical system with that of sending power along good metallic conductors, there is at once apparent a very serious objection in the needless dispersion of energy throughout space in every direction. If a power generator by wireless transmission, without any metallic connection, can work one motor at a distance of, say, 1,000 miles, then it can also operate millions of similar possible motors situated at the same distance; and by far the greater part of its electro-motive force must be wasted in upward dispersion.

The analogy of the wireless transmitter of intelligence may be misleading if applied to the question of power. The practicability of wireless telegraphy depends upon the marvellous susceptibility of the "coherer," which enables it to respond to an impulse almost infinitesimally small, certainly very much smaller than that despatched by the generator from the receiving station. From this it follows, as already stated, that the analogy of apparatus designed merely for the despatch of intelligence by signalling cannot safely be applied to the case of the transmission of energy.

Making all due allowances for the prospects of advance in minimising the resistance of the atmosphere, it must nevertheless be remembered that any wireless system will be called upon to compete with improved means of conveying the electric current along metallic circuits. Electrical science, moreover, is only at the commencement of its work in economising the cost of power-cables.

The invention by which one wire can be used to convey the return current of two cables very much larger in sectional area is only one instance in point. The two major cables carry currents running in opposite directions, and as these currents are both caused to return along the third and smaller wire their electro-motive forces balance one another, with the result that the return wire needs only to carry a small difference-current. The return wire, in fact, is analogous to the Banking Clearing House, which deals with balances only, and which therefore can sometimes adjust business to the value of many millions with payments of only a few thousands. Later on it may fairly be expected that duplicate and quadruplicate telegraphy will find its counterpart in systems by which different series of electrical impulses of high voltage will run along a wire, the one alternating with the other and each series filling up the gaps left between the others.

CHAPTER IV
ARTIFICIAL POWER

The steam-turbine is the most clearly visible of the revolutionary agencies in motors using the artificial sources of power. In the first attempts to introduce the principle the false analogy of the water-turbine gave rise to much waste of inventive energy and of money; but the more recent and more distinctly successful types of machine have been constructed with a clear understanding that the windmill is the true precursor of the steam-turbine. It is clearly perceived that, although it may be convenient and even essential to reduce the arms to pigmy dimensions and to enclose them in a tube, still the general principle of the machine must resemble that of a number of wind motors all running on the same shaft.

It has been proved, moreover, that this multiplicity of minute wheels and arms has a very distinct advantage in that it renders possible the utilisation of the expansive power of steam. The first impact is small in area but intense in force, while those arms which receive the expanded steam further on are larger in size as suited to making the best use of a weaker force distributed over a greater amount of space.

The enormous speed at which steam under heavy pressure rushes out of an orifice was not duly appreciated by the first experimenters in this direction. To obtain the best results in utilising the power from escaping steam there must be a certain definite proportion between the speed of the vapour and that of the vane or arm against which it strikes. In other words, the latter must not "smash" the jet, but must run along with it. In the case of the windmill the ratio has been stated approximately by the generalisation that the velocity of the tips of the sails is about two and a half times that of the wind. This refers to the old style of windmill as used for grinding corn.

The steam turbine must, therefore, be essentially a motor of very great initial speed; and the efforts of recent inventors have been wisely directed in the first instance to the object of applying it to those purposes for which machinery could be coupled up to the motor with little, if any, necessity for slowing down the motion through such appliances as belting, toothed wheels, or other forms of intermediate gearing. The dynamo for electric lighting naturally first suggested itself; but even in this application it was found necessary to adopt a rate of speed considerably lower than that which the steam imparts to the turbine; and, unfortunately, it is exactly in the arrangement of the gear for the first slowing-down that the main difficulty comes in.

Nearly parallel is the case of the cream separator, to which the steam-turbine principle has been applied with a certain degree of success. By means of fine flexible steel shafts running in bearings swathed in oil it has been found possible to utilise the comparatively feeble force of a small steam jet operating at immense speed to produce one of much slower rate but enormously greater strength. Some success has been achieved also in using the principle not only for cream separators, which require a comparatively high velocity, but for other purposes connected with the rural and manufacturing industries.

An immense forward stride, however, was made when the idea was first conceived of a steam-turbine and a water-turbine being fixed on the same shaft and the latter being used for the propulsion of a vessel at sea. In this case it is obvious that, by a suitable adjustment of the pitch of screw adopted in both cases, a nice mathematical agreement between the vapour power and the liquid application of that power can be ensured.

All previous records of speed have been eclipsed by the turbine-driven steamers engined on this principle. Through the abolition of the principal causes of excessive vibration—which renders dangerous the enlargement of marine reciprocating engines beyond a certain size—the final limit of possible speed has been indefinitely extended. The comfort of the passenger, equally with the safety of the hull, demands the diminution of the vibration nuisance in modern steamships, and whether the first attempts to cater for the need by turbine-engines be fully successful or not, there is no doubt whatever that the fast mail packets of the future will be driven by steam-engines constructed on a system in which the turbine principle will form an important part.

Further applications will soon follow. It is clear that if the steam-turbine can be advantageously used for the driving of a vessel through the water, then, conversely, it can be similarly applied to the creation of a current of water or of any other suitable liquid. This liquid-current, again, is applicable to the driving of machinery at any rate that may be desired. In this view the slowing-down process, which involves elaborate and delicate machinery when accomplished in the purely mechanical method, can be much more economically effected through the friction of fluid particles.

One method of achieving this object is an arrangement in which the escaping steam drives a turbine-shaft running through a long tube and passing into the water in a circular tank, in which, again, the shaft carries a spiral or turbine screw for propelling the water. The arrangement, it will be seen, is strictly analogous to that of the steam-turbine as used in marine propulsion, the shaft passing through the side of the tank just as it does through the stern of the vessel.

One essential point, however, is that the line of the shaft must not pass through the centre of the circular tank, but must form the chord of an arc, so that when the water is driven against the side by the revolution of the screw it acts like a tangential jet. Practically the water is thus kept in motion just as it would be if a hose with a strong jet of water were inserted and caused to play at an obtuse angle against the inner side.

Motion having been imparted to the fluid in the tank, a simple device such as a paddle-wheel immersed at its lower end, may be adopted for taking up the power and passing it on to the machinery required to be actuated. By setting both the shaft carrying the vanes for the steam-turbine and the screw for the propulsion of the water at a downward inclination it becomes practicable to drive the fluid without requiring any hole in the tank; and in this case the latter may be shaped in annular form and pivoted so that it becomes a horizontal fly-wheel. Obstructing projections on the inside periphery of the annular tank assist the water to carry the latter along with it in its circular motion.

For small steam motors, particularly for agricultural and domestic purposes, the turbine principle is destined to render services of the utmost importance. The prospect of its extremely economical construction depends largely upon the fact that, with the exception of two or three very small bearings carrying narrow shafts, it contains no parts demanding the same fine finish as does the cylinder of a reciprocating engine. It solves in a very simple manner the much-vexed problem of the rotary engine, upon which so much ingenuity has been fruitlessly exercised. The steam-turbine also has shown that, for taking advantage of the generation and the expansive power of steam, there is no absolute necessity for including a steam-tight chamber with moving parts in the machine.

For very small motors suitable for working fans and working other household appliances, the use of a jet of steam, applied directly to drive a small annular fly-wheel filled with mercury—without the intervention of any turbine—will no doubt prove handy. But in the economy of the future such appliances will take the place of electrical machinery only in exceptional situations.

One promising use of the turbine or steam-jet—used to propel a fly-wheel filled with liquid as described—has for its object the supply of the electric light in country houses. In this case the fly-wheel is fitted, on its lower side, to act as the armature of a dynamo, and the magnets are placed horizontally around it.

The full effective power from a jet of steam is not communicated to a dynamo for electric lighting or other purposes unless there be a definite ratio between the speeds of the turbine and of the armature respectively. This may be conveniently provided for, with more precision and in a less elaborate way than that which has just been described, if the steam jet be made to drive a vertically pendant turbine, the lower extremity of which, carrying very small horizontal paddles, must be inserted into the centre of a circular tank.

The principle upon which the reduction of speed necessary for the dynamo is then effected depends upon the fact that in a whirlpool the liquid near the centre runs nearly as fast as that on the outer periphery, and therefore—the circles being so very much smaller—the number of revolutions effected in a given time is much greater. Thus a steam jet turning a pendant turbine—dipping into the middle of the whirlpool and carrying paddles—at an enormously high speed may be made to impart motion to the water in a circular tank (or, if desired, to the tank itself) at a very much slower rate; the amount of the reduction, of course, depending mainly on the ratio between the diameter of the tank and the length of the small paddles at the centre setting the liquid in motion.

For special purposes it is best to substitute a spherical for an ordinary circular tank and the size may be greatly diminished by using mercury instead of water. The sphere is complete, excepting for a small aperture at the top for the admission of the steel shaft of the steam-driven turbine. No matter how high may be the speed, the liquid cannot be thrown out from a spherical revolving receptacle constructed in this way. Moreover, the mercury acts not only as a transmitter of the power from the turbine to the purpose for which it is wanted, but also as a governor. Whenever the speed becomes so great as to throw the liquid entirely into the sides of the sphere—so that the shaft and paddles are running free of contact with it in the middle—the machine slows down, and it cannot again attain full speed until the same conditions recur.

The rate of speed which may be worked up to as a maximum is determined by the position of the paddle-wheel, which is adjustable and floats upon the liquid although controlled in its circular motion by the shaft which passes through a square aperture in it and also a sleeve extending upward from it. The duty of the latter is to economise steam by cutting off the jet as soon as, by its rapidity of motion, the paddle-wheel has thrown the mercury to the sides to such an extent as to sink to a certain level in the centre.

Cheap motors coupled with cheap dynamos will, in the twentieth century, go far towards lightening the labours of millions whose toil is at present far too much of a mere mechanical nature. The dynamo itself, however, requires to be greatly reduced in first cost. Particularly it is necessary that the expense involved in drawing the wire, insulating it, and winding machines with it, should be diminished. This will no doubt be partly accomplished by the electrolytic producers of copper when once they get properly started on methods of depositing thin strips or wires of tough copper on to sheets of insulating material for wrapping round the magnets and other effective parts intended for dynamos. There is no fundamental reason which forbids that when electro deposition is resorted to for the recovery of a metal from its ore it should be straightway converted to the shape and to the purpose for which it is ultimately intended. This consideration has presented itself to the minds of some of the manufacturers of aluminium, who make many articles intended for household use electrolytically; and it must affect many other trades which are concerned in the output and in the working-up of metals readily susceptible of deposition—more particularly such as copper.

The familiar aneroid barometer furnishes a hint for another convenient form of small steam-engine. In seeking to cheapen machinery of this class it is of the utmost importance that the necessity for boring out cylinders and for planing and other expensive work should be avoided. In the aneroid barometer a shallow circular box is fitted with a cover, which is corrugated in concentric circles, and the pressure of the superincumbent air is caused to depress the centre of this cover through the device of partially exhausting the box of air and thus diminishing the internal resistance. To the slightly moving middle part of the cover is affixed a lever which actuates, after some intermediate action, the hand which moves on the dial to indicate, by its record of variations in the weight of the atmosphere, what the prospect of the weather may be.

In the aneroid form of the steam-engine the cylinder is immensely widened and flattened, and the broad circular lid, with its spiral corrugations, takes the place of the piston. The rod, which acts virtually as a piston-rod, is hollow, and it works into a bearing which permits the steam to escape when the extreme point of the stroke has been reached into a separate condensing chamber kept cool with water. The boiler itself, with corrugated top, may take the place of the cylinder.

In some respects this little machine represents a retrograde movement, even from Watt's original engine with its separate condenser; but its extreme economy of first cost recommends it to poor producers. In the near future no country homestead will be without its power installation of one kind or another, and there is room for many types of cheap motors.

A motor like the steam-turbine is evidently the forerunner of other engines designed to utilise the force of an emission jet of vapour or gas. There are very many processes in which gases generated by chemical combinations are permitted to escape without performing any services, not even that of giving up the energy which they may be made to store up when held in compression in a closed vessel.

The reciprocating forms found suitable for steam and gas engines are hardly adaptable for experiments in the direction of economising this source of power, one fatal objection in the majority of cases being the corrosive effects of the gases generated upon the insides of cylinders and other working parts. As soon as the force of the emission jet can be applied as a factor in giving motive power, the fact that no close-fitting parts are required for the places upon which the line of force impinges will alter the conditions of the whole problem. In the centrifugal sand pump, as now largely used for raising silt from rivers and harbours, the serious corrosive action of the jet of sand and water upon the inside of the pump has been successfully overcome by facing the metal with indiarubber; but nothing of the kind could have been done if the working of the apparatus had depended on the motion of close-fitting parts, as in the ordinary suction or lift pump.

As an instance of the class of work for which gaseous jets, for driving turbines or similar forms of motor, may perform useful services the case of farm-made superphosphate of lime may be cited. By subjecting bones to the action of sulphuric acid the farmer may manufacture his own phosphatic manures for the enrichment of his land. But the carbonic dioxide and other gases generated as the result of the operation are wasted. Therefore it at present pays better to carry the bones to the sulphuric acid than to reverse the procedure by conveying the acid to the farm, where the bones are a by-product.

So bulky are the latter, however, that serious waste of labour is involved in transporting them for long distances. Calculations made out by the experts of various state agricultural stations show that, as a general rule, it is now cheaper for the farmer to buy his superphosphates ready made than to make them on his farm. The difference in some cases, however, is not great; and only a comparative trifle would be needed in order to turn the balance. This may probably be found in the economic value of the service rendered by a turbine-engine or other device for utilising the expansive power of the gases which are driven from the constituents of the bones by the action of the sulphuric acid.

For pumping water and other ordinary farm operations the chemical gas-engine will prove very handy; and the great point in its favour will be that instead of useless cinders the refuse from it will consist of the most valuable compost with which the farmer can dress the soil. Enamelled iron will be employed for the troughs in which the bones and acid will be mixed, and a cover similar to that placed over a "Papin's digester" will be clamped to the rim all round, the gases being liberated only in the form of a jet used for driving machinery.

For very small motors, applicable specially to domestic purposes such as ventilation, there is one source of power which, in all places within the reticulation areas of waterworks, may be had practically for nothing. Probably when the owners of water-supply works realise that they have command of something which is of commercial value, although hitherto unnoticed, they will arrange to sell not only the water which they supply, but also the power which can be generated by its escape when utilised and by the variations in the pressure from hour to hour and even from minute to minute.

The latter, for such purposes as ventilation, for instance, will no doubt come to the front sooner than the intermittent power now wasted by the outflowing of water—a power which is comparatively too small an item in most cases to compensate for the outlay and trouble of arranging for the storage of energy. But in the case of the variation in the pressure, without any escape of water at all, no such disability appears. Experiments conducted in several of the larger cities of England with various types of water meters—which are really motors on a small scale—have proved the practicability of obtaining a source of constant power from what may be termed the ebb and the flow of pressure within the pipes of a water supply system.

At every hour of the day there is a marked variation in the quantity of water that is being drawn away by consumers, and consequently a rise and fall in the degree of pressure recorded by the meter. In an apparatus for converting the power derivable from this source to useful purposes something on a very small scale analogous to that which has already been described in connection with utilising the rise and fall of a wave will be found serviceable. A small spur-wheel is gripped on two sides by two metal laths, with edges serrated like those of saws, and held against the wheel by gentle pressure. Every movement of the two saws—whether backwards or forwards—is then responded to by a continuous circular motion of the wheel, with the sole exception of those movements which may be too small in extent to include even as much as a single tooth of the wheel. On this account it is important that the teeth should be made as numerous as possible consistently with the amount of pressure which they may have to bear.

Resort may be had to the principle of the aneroid barometer in order to secure from the water within the pipe-system the energy by which these saw-like bands are driven up and down with reciprocal motion. A very shallow circular tank in the shape of a watch is in communication with the water in the pipes, and its top or covering is composed of a concentrically-corrugated sheet of finely tempered steel. At the centre of this is fixed the guide which pushes and pulls the saw-like laths. Every rise and fall in the pressure of the water now effects a movement of the spur-wheel, and the latter may conveniently be connected with the strong spring of a clockwork attachment, so that the water pressure is really used for winding up a clockwork ventilating-fan.

In the making of cheap steam and gas engines, as well as in machine work generally, rapid progress will be made when the possibilities of producing hard and smooth wearing surfaces without the need for cutting and filing rough-cast metal have been fully investigated. Many parts of machinery will be electro-deposited—like the small articles already mentioned—in aluminium or hard copper at the metallurgical works where ore is being treated for the recovery of metal, or even at the mines themselves.

Side by side with this movement there will be one for developing the system of stamping mild steel and then tempering it. At the same time also the behaviour of various metals and alloys, not only in the cold state but also at the critical point between melting and solidification, will be much more carefully studied so as to take advantage of every means whereby accurately shaped articles may be made and finished in the casting. It has been found, for example, that certain kinds of type metal, if placed under very heavy pressure at the moment when passing from the liquid to the solid condition, not only take the exact form of the mould in which they are placed, but become extremely hard by comparison with the same alloy if permitted to solidify without pressure.

The example of the cheap watch industry may be cited to convey an idea of the immensely important revolution which will take place in the production of both small and large prime-motors when all the possibilities of electrotyping, casting, and stamping the various wearing parts true to shape and size have been fully exploited. An accurate timekeeper is now practically within the reach of all; and in the twentieth century no one who requires a small prime motor to do the rough work about home or farm will be compelled to do without it by reason of poverty—unless, perhaps, he is absolutely destitute and a fit subject for public charity.

Many domestic industries which were crushed out of existence during the early part of the nineteenth century will therefore be resuscitated. The dear steam-engine created the factory system and brought the operatives to live close together in long rows of unsightly dwellings, but the cheap engine, in conjunction with the motor driven by transmitted electricity, will give to the working people comparative freedom again to live where they please, and to enjoy the legitimate pleasures both of town and of country.

CHAPTER V
ROAD AND RAIL

The existing keen motor-car rivalry presents one of the most interesting and instructive mechanical problems which are left still unsolved by the close of the nineteenth century. The question to be determined is not so much whether road locomotion by means of mechanical power is practicable and useful, for, of course, that point has been settled long ago; indeed it would have been recognised as settled years before had it not been for the crass legislation of a quarter of a century since which deliberately drove the first steam-motors off the road in order to ensure the undisturbed supremacy of horse traffic. The real point at issue is whether a motor can be made which shall furnish power for purposes of road locomotion as cheaply and conveniently as is already done for stationary purposes.

Horse traction, although extremely dear, possesses one qualification which until the present day has enabled it to outdistance its mechanical competitors upon ordinary roads. This is its power of adapting itself, by special effort, to the exigencies caused by the varying nature of the road. Watch a team of horses pulling a waggon along an undulating highway, with level stretches of easy going and here and there a decline or a steep hill. There is a continual adjustment of the strain which each animal puts upon itself according to the character of the difficulties which must be surmounted, the effort varying from nothing at all—when going down a gentle decline—up to the almost desperate jerk with which the vehicle is taken over some stony part right on the brow of an eminence. The whip cracks and by threats and encouragements the driver induces each horse to put forth, for one brief moment, an effort which could not be sustained for many minutes save at the peril of utter exhaustion.

When the unit of nominal horse-power was fixed at 33,000 foot-pounds per minute the work contemplated in the arbitrary standard was supposed to be such as a horse could go on performing for several hours. It was, of course, well recognised that any good, upstanding horse, if urged to a special effort, could perform several times the indicated amount of work in a minute.

Nevertheless the habit of reckoning steam-power in terms of a unit drawn from the analogy of the horse undoubtedly tended for many years to obscure the essential difference between the natures of the two sources of power. Railroads were built with the object of rendering as uniform as possible the amount of power required to transport a given weight of goods or passengers over a specified distance; and consequently the application of the steam-engine to traffic conducted on the railway line was a success. Many inventors at once jumped to the conclusion that, by making some fixed allowance for the greater roughness of an ordinary road, they would be able to construct a steam-traction engine that would suit exactly for road traffic. In a rough and rudimentary way an attempt to provide for the special effort required at steep or stony places was made by the introduction of a kind of fly-wheel of extraordinary weight proportionate to the size of the engine; and the same object was aimed at by increasing the power of the engine to somewhere near the limit of the possible special requirements. The consequence was the evolution of an immensely ponderous and wasteful machine, which for some years only held its ground within the domain of the heavy work of roadmaking. As a means of road traction the steam-engine was for half a century almost entirely discomfited and routed by horse-power, partly owing to this mechanical defect and partly, as we have seen, through legislative partisanship.

The explosive type of engine was next called into requisition to do battle against the living competitor of the engineer's handiwork. Petroleum and alcohol, when volatilised and mixed with air in due proportion, form explosive mixtures which are much more nearly instantaneous in their action than an elastic vapour like steam held under pressure in a boiler, and liberated to perform its work by comparatively slow expansion. The petroleum engine, as applied to the automobile, does its work in a series of jerks which provide for the unequal degrees of power required to cope with the unevenness of a road.

As against this, however, there are certain grave defects, due mainly to the use of highly inflammable oils vapourised at high temperatures; and these have impressed a large proportion of engineers with a belief that, in the long run, either electricity or steam will win the day. Storage batteries are well adapted for meeting the exigencies of the road, just as they are for those of tramway traffic, because, as soon as an extra strain is to be met, there is always the resource of coupling up fresh batteries held in reserve—a process which amounts to the same as yoking new horses to the vehicle in order to take it up a hill. In practice, however, it is found that the jerky vibratory motion of the gasoline automobile provides for this in a way almost as convenient, although not so pleasant.

The chance of the steam-engine being largely adopted for automobile work and for road traffic generally depends principally on the prospects of inventing a form of cylinder—or its equivalent—which will enable the driver to couple up fresh effective working parts of his machinery at will, just as may be done with storage batteries. A new form of steam cylinder designed to provide for this need will outwardly resemble a long pipe— one being fixed on each lower side of the vehicle—but inwardly it will be divided into compartments each of which will have its own separate piston. Practically there will thus be a series of cylinders having one piston-rod running through them all, but each having its own piston.

Normally, this machine will run with an admission of steam to only one or two of the cylinders; but when extra work has to be done the other cylinders will be called into requisition by the opening of the steam valves leading to them. Provision can be made for the automatic working of this adjustment by the introduction of a spring upon the piston-rod, so arranged that, as soon as the resistance reaches a certain point, a lever is actuated which opens the valves to admit steam to the reserve cylinders of the engine. On such occasions, of course, the consumption of steam must necessarily be greatly increased; but on the other hand the automatic system of the admission to each cylinder also results in a shutting off of the steam when little or no work is required. In fact, with a fully automatic action, regulating the consumption of steam exactly according to the amount of force necessary to drive the automobile, it would be possible to work even a single cylinder to much greater advantage than is done by the machines generally in use.

So heavy are the storage batteries needed for electric traction of the road motor-car that practically it is not found convenient to carry enough of cells to last for more than a twenty-mile run. The batteries must then either be replaced, or a delay of some three hours must occur while they are being recharged. The idea of establishing charging stations at almost every conceivable terminus of a run is quite chimerical; and, even if hundreds of such stations were provided for the convenience of the users of electric traction, the limitation imposed by being forced to follow the established routes would always give to the non-electric motor an advantage over its competitor.

The best hope for the storage battery on the automobile rests upon its convenience as a repository of reserve power in conjunction with such a prime motor as the steam-engine. A turbine worked by a jet of steam, as already described, and moving in a magnetic field to generate electricity for storage in a few cells, is a convenient form in which steam and electricity can be yoked together in order to secure a power of just the type suitable for driving an automobile. In the machine indicated the supply of the motive power is direct from the storage batteries, which can be coupled up in any required number according to the exigencies of the road. Automatic gear may be introduced by an adaptation of the principle already referred to.

In a light road-motor for carrying one or two persons on holiday trips or business rounds, the quality of adaptability of the source of power to the sudden demands due to differences of level in the road is not so absolutely essential as it is in traction engines designed for the transport of goods over ordinary roads. In the former class of work the waste of power involved in employing a motor of strength sufficient to climb hills—although the bulk of the distance to be travelled is along level roads—may not be at all so serious as to overbalance the many and manifest advantages of the automobile principle. At the same time, as has already been indicated, there is no doubt whatever that when proper automatic shut-off contrivances have been applied for economising mechanical energy in the passenger road-motor, an immense impetus will be given to its advancement.

In the road traction-engine the need for what may be termed *effort* on the part of the mechanism is much greater, more especially as the competition against horse-traction is conducted on terms so much more nearly level. A team of strong draught-horses driven by one man on a well-loaded waggon is a far more economical installation of power than a two-horse buggy carrying one or two passengers.

The asphalt and macadamised tracks which are now being laid down along the sides of roads for the convenience of cyclists, are the significant forerunners of an improvement destined to produce a revolution in road traffic during the twentieth century. When automobiles have become very much more numerous, and local authorities find that the settlement of wealthy or comparatively well-to-do families in their neighbourhoods may depend very largely upon the question whether light road-motor traffic may be conveniently conducted to and from the nearest city, an immense impetus will be administered to the reasonable efforts made for catering for the demand for tracks for the accommodation of automobiles, both private and public.

The tyranny of the railway station will then be to a large extent mitigated, and suburban or country residents will no longer be practically compelled to crowd up close to each station on their lines of railroad. Under existing conditions many of those who travel fifteen or twenty miles to business every day live just as close to one another, and with nearly as marked a lack of space for lawn and garden, as if they lived within the city. The bunchy nature of settlement promoted by railways must have excited the notice of any intelligent observer during the past twenty or thirty years—that is to say since the suburban railroad began to take its place as an important factor in determining the locating of population.

To a very large extent the automobile will be rather a feeder to the railway than a rival to it; and all sorts of by-roads and country lanes will be improved and adapted so as to admit of residents running into their stations by their own motor-cars and then completing their journeys by rail. But when this point has been reached, and when fairly smooth tracks adapted for automobile and cycling traffic have been laid down all over the country, a very interesting question will crop up having reference to the practicability of converting these tracks into highways combining the capabilities both of roads and of railways.

In an ordinary railroad the functions of the iron or steel rails are twofold, first to carry the weight of the load, and second to guide the engine, carriage or truck in the right direction. Now the latter purpose—in the case of a rail-track never used for high speeds, especially in going round curves—might be served by the adoption of a very much lighter weight of rail, if only the carrying of the load could be otherwise provided for. In fact, if pneumatic-tyre wheels, running on a fairly smooth asphalt track, were employed to bear the weight of a vehicle, there would then be no need for more than one guide-rail, which might readily be fixed in the middle of the track; but this should preferably be made to resemble the rail of a tram rather than that of a railroad.

"Every man his own engine-driver" will be a rule which will undoubtedly require some little social and mechanical adjustment to carry out within the limits of the public safety. But the automobile, even in its existing form, makes the task of completing this adjustment practically a certainty of the near future; and as soon as it is seen that motor tracks with guide lines render traffic safer than it is on ordinary roads, the main objections to the innovation will be rapidly overcome. The rule of the road for such guide-line tracks will probably be based very closely on that which at present exists for ordinary thoroughfares. On those roads where two tracks have been laid down each motor will be required to keep to the left, and when a traveller coming up behind is impatient at the slow rate of speed adopted by his precursor he will be compelled to make the necessary détour himself, passing into the middle of the thoroughfare and there outstripping the party in front, without the assistance of the guide-rail, and rejoining the track.

To execute this movement, of course, the motor wheels for the guide-tracks must be mounted on entirely different principles from those adapted for railroad traffic. The broad and soft tyred wheels which bear upon the asphalt track will be entrusted with the duty of carrying the machine without extraneous aid; but there will be two extra wheels, one in front and one at the rear, capable of being lifted at any time by means of a lever controlled by the driver. These guiding wheels will fit into the groove of the tram line in the centre, being made of a shape suitable for enabling the driver to pick up the groove quickly whenever he pleases. The carrying wheels of the vehicle in this system are enabled to pass over the guide-rail readily, because the latter does not stand up from the track like the line in a railroad.

A simpler plan, particularly adapted for roads which are to have only a single guide-rail, is to place the rail at the off-side of the track, and to raise it a few inches from the ground. The wheels for the rail are attached to arms which can be raised and lifted off the rail by the driver operating a lever. Guiding irons, forming an inverted Y, are placed below the bearings of the wheels to facilitate the picking up of the rail, their effect being that, if the driver places his vehicle in approximately the position for engaging the side wheels with the rail and then goes slowly ahead, he will very quickly be drawn into the correct alignment. Of course the rails for this kind of track can be very light and inexpensive in comparison with those required for railroads on which the whole weight of each vehicle, as well as the lateral strain caused by its guidance, must fall upon the rail itself.

The asphalt track and its equivalent will be the means of bringing much nearer to fulfilment the dream of having "a railway to every man's door". Many such tracks will be equipped with electric cables as well as guiding-rails, so that cars with electric motors will be available for running on them, and the power will be supplied from a publicly-maintained station. Some difficulty may at first be experienced in adjusting the rates and modes of payment for the facilities thus offered; but a convenient precedent is present to hand in the class of enactment under which tramway companies are at present protected from having their permanent ways used by vehicles owned by other persons. Practically the possession of a vehicle having a flanged wheel and a gauge exactly the same as that of the tram lines in the vicinity may be taken to indicate an intention to use the lines. Similarly a certain relation between the positions of guiding wheels and those of the connections with cables may be held to furnish evidence of liability to contribute towards the maintenance of motor-tracks.

Roads and railways will be much more closely inter-related in the future than they have been in the past. The competition of the automobile would in itself be practically sufficient to force the owners of railways into a more adaptive mood in regard to the true relations between the world's great highways. The way in which the course of evolution will work the problem out may be indicated thus:—First, the owners of automobiles will find it convenient in many instances to run by road to the nearest railway station which suits their purposes, leaving their machines in charge of the stationmaster and going on by train. In course of time the owners of "omnibus automobiles" will desire to secure the same advantage for their customers, and on this account the road cars will await the arrival and departure of every train just as horse vehicles do at present. The next step will be taken by the railway companies, or by the local authorities, when it becomes obvious that there is much more profit in motor traffic than there ever was in catering for the public by means of vehicles drawn by horses. Each important railway station will have its diverging lines of motor-traffic for the convenience of passengers, some of them owned and managed by the same authority as the railway line itself.

Rivalry will shortly enforce an improvement upon this system, because in the keen competition between railway lines those stations will attract the best parts of the trade at which the passengers are put to the smallest amount of inconvenience. The necessity for changing trains, with its attendant bustle of looking after luggage, perhaps during very inclement weather, always acts as a hindrance to the popularity of a line. When "motor-omnibuses" are running by road all the way into the city, setting people down almost at their doors and making wide circuits by road, the proprietors of these vehicles will make the most of their advantages in offering to travellers a cosy and comfortable retreat during the whole of their journey.

Road-motors, comfortably furnished, will therefore be mounted upon low railway trucks of special construction, designed to permit of their being run on and off the trucks from the level of the ground. The plan of mounting a road vehicle upon a truck suited to receive it has already been adopted for some purposes, notably for the removal of furniture and similar goods; and it is capable of immense extension. An express train will run through on the leading routes from which roads branch out in all directions, and as it approaches each station it will uncouple the truck and "motor-omnibus" intended for that destination. The latter will be shunted on to a loopline. The road-motor will be set free from its truck and will then proceed on its journey by road.

When a similar system has been fully adapted for the conveyance of goods by rail and road experiments will then be commenced, on a systematic basis, with the object of rendering possible the picking up of packages, and even of vehicles, without stopping the train. The most pressing problem which now awaits solution in the railway world is how to serve roadside stations by express trains. "Through" passengers demand a rapid service; while the roadside traffic goes largely to the line that offers the most frequent trains. In the violent strain and effort to combine these two desiderata the most successful means yet adopted have been those which rely upon the destruction of enormous quantities of costly engine-power by means of quick-acting brakes. The amount of power daily converted into the mischievous heat of friction by the brakes on some lines of railway would suffice to work the whole of the traffic several times over; but the sacrifice has been enforced by the public demand for a train that shall run fast and shall yet stop as frequently as possible.

Progress in this direction has reached its limit. A brake that shall conserve, instead of destroying, the power of the train's inertia on pulling up at a station is urgently required; but the efforts towards supplying the want have not, so far, proved very successful. Each carriage or truck must be fitted with an air-pump so arranged that, on the application of the brake by the engine-driver, it shall drive back a corresponding amount of air to that which has been liberated from the reservoir, and the energy thus stored must be rendered available for re-starting the train. Trials in this direction have been made through the application of strong springs which are caused to engage upon the wheels when the brake is applied, and thus are wound up, but which may then be reversed in position, so that for the starting of the vehicle the rebound of the spring offers material assistance. It is obvious, however, that the use of compressed air harmonises better with the railway system than any plan depending upon springs. The potential elasticity in an air-reservoir of portable dimensions is enormously greater than that of any metallic spring which could conveniently be carried.

In picking up and setting down mail-bags a system has been for some years in operation on certain railway lines indicating in a small way the possibilities of the future in the direction of obviating the need for stopping trains at stations. The bag is hung on a sliding rod outside of the platform, and on a corresponding part of the van is affixed a strong net, which comes in contact with the bag and catches it while the train goes past at full speed. Dropping a bag is, of course, a simpler matter.

The occasionally urgent demand for the sending of parcels in a similar manner has set many inventive brains to work on the problem of extending the possibilities of this system, and there seems no reason to doubt that before long it will be practicable to load some classes of small, and not readily broken, articles into trucks or vans while trains are in motion.

The root idea from which such an invention will spring may be borrowed from the sliding rail and tobogganing devices already introduced in pleasure grounds for the amusement of those who enjoy trying every novel excitement. A light and very small truck may be caused to run down an incline and to throw itself into one of the trucks comprising a goods train. The method of timing the descent, of course, will only be definitely ascertained after careful calculation and experiments designed to determine what length of time must elapse between the liberation of the small descending truck and the passing of the vehicle into which its contents are to be projected.

Foot-bridges over railway lines at wayside stations will afford the first conveniences to serve as tentative appliances for the purpose indicated. From the overway of the bridge are built out two light frameworks carrying small tram-lines which are set at sharp declivities in the directions of the up and the down trains respectively, and which terminate at a point just high enough to clear the smoke-stack of the engine.

The small truck, into which the goods to be loaded are stowed with suitable packings to prevent undue concussion, is held at the top of its course by a catch, readily released by pressure on a lever from below. The guard's van is provided at its front end with a steel, upright rod carrying a cross-piece, which is easily elevated by the guard or his assistant in anticipation of passing any station where parcels are to be received by projection. At the rear of the van is an open receptacle communicating by a door or window with the van itself. At the instant when the steel cross-piece comes in contact with the lever of the catch, which holds the little truck in position on the elevated footbridge, the descent begins, and by the time that the receptacle behind the van has come directly under the end of the sloping track the truck has reached the latter point and is brought to a sudden standstill by buffers at the termination of the miniature "toboggan". The ends of the little truck being left open, its contents are discharged into the receptacle behind the van, from which the guard or assistant in charge removes them into the vehicle itself. For catching the parcels thrown out from the van a much simpler set of apparatus is sufficient.

On a larger scale, no doubt in course of time, a somewhat similar plan will be brought into operation for causing loaded trucks to run from elevated sidings and to join themselves on to trains in motion. One essential condition for the attainment of this object is that the rails of the siding should be set at such a steep declivity that, when the last van of the passing train has cleared the points and set the waiting truck in motion by liberating its catch, the rate of speed attained by the pursuing vehicle should be sufficiently high to enable it to catch the train by its own impetus.

It may be found more convenient on some lines to provide nearly level sidings and to impart the necessary momentum to the waiting truck, partly through the propelling agency of compressed air. Any project for what will be described as "shooting a truck loaded with valuable goods after the retreating end of a train," in order to cause it to catch up with the moving vehicles, will no doubt give rise to alarm; and this feeling will be intensified when further proposals for projecting carriages full of passengers in a similar method come up for discussion. But these apprehensions will be met and answered in the light of the fact that in the earlier part of the nineteenth century critics of what was called "Stephenson's mad scheme" of making trains run twenty or even thirty miles an hour were gradually induced to calm their nerves sufficiently to try the new experience of a train journey!

The wire-rope tramway has hitherto been used principally in connection with mines situated in very hilly localities. Trestles are erected at intervals upon which a strong steel rope is stretched and this carries the buckets or trucks slung on pulley-blocks, contrived so as to pass the supports without interference. A system of this kind can be worked electrically, the wire-rope being employed also for the conveyance of the current. But an inherent defect in the principle lies in the fact that the wire-rope dips deeply when the weight passes over it, and thus the progress from one support to another resolves itself into a series of sharp descents, followed by equally sharp ascents up a corresponding incline. The usual way of working the traffic is to haul the freight by means of a rope wound round a windlass driven by a stationary engine at the end. The constantly varying strain on the cable proves how large is the amount of power that must be wasted in jerking the buckets up one incline to let them jolt down another when the point of support has been passed.

Hitherto the wire-rope tramway has been usually adopted merely as presenting the lesser of two evils. If the nature of the hills to be traversed be so precipitous that ruinous cuttings and bridges would be needed for the construction of an ordinary railway or tramway line, the idea of conveyance by wire suggests itself as being, at least, a temporary mode of getting over the difficulty. But a great extension of the principle of overhead haulage may be expected as soon as the dipping of the load has been obviated, and the portion of the moving line upon which it is situated has been made rigid. A strong but light steel framework, placed in the line of the drawing-cable, and of sufficient length to reach across two of the intervals between the supports, may be drawn over enlarged pulleys and remain quite rigid all the time.

The weight-carrying wire-rope is thus dispensed with, and the installation acquires a new character, becoming, in point of fact, a moving bridge which is drawn across its supports and fits into the grooves in the wheels surmounting the latter. The carriage or truck may be constructed on the plan adopted for the building of the longest type of modern bogie carriages for ordinary railways, the tensile strength of steel rods being largely utilised for imparting rigidity. We now find that instead of a railway we have the idea of what may be more appropriately called a "wheelway". The primitive application of the same principle is to be seen in the devices used in dockyards and workshops for moving heavy weights along the ground by skidding them on rollers. Practically the main precaution observed in carrying out this operation is the taking care that no two rollers are put so far apart that the centre of gravity of the object to be conveyed shall have passed over one before the end has come in contact with the next just ahead of it.

The "wheelway" itself will be economical in proportion as the length of the rigid carriage or truck which runs upon it is increased. The carrying of cheap freight will be the special province of the apparatus, and it will therefore be an object to secure the form of truck which will give, with the least expense, the greatest degree of rigidity over the longest stretch of span from one support to another. Some modification of the tubular principle will probably supply the most promising form for the purpose. The hope of this will be greatly enhanced through the recent advances in the art of tube-constructing by which wrought-iron and tough steel tubes can be made quite seamless and jointless, being practically forged at one operation in the required tubular shape.

For mining and other similar purposes, the long tubal "wheelway" trucks of this description can be drawn up an incline at the loading station so as to be partially "up-ended" in position for receiving the charges or loads of mineral or other freight. After this they can be despatched along the "wheelway" on the closing of the door at the loading end. In regard to the mode of application of the power in traction, the shorter-distance lines may serve their objects well enough by adopting the endless wire-rope system at present used on many mining properties.

But it is found in practice that for heavy freight this endless cable traction does not suit over distances of more than about two miles. Mining men insist upon the caution that where this length of distance has to be exceeded in the haulage of ore from the mine over wire-rope tramways, there is need for two installations, the loaded trucks being passed along from one to the other by means of suitable appliances at the termini.

Electric traction must, in the near future, displace such a cumbrous system, and the plan upon which it will be applied will probably depend upon the use of a steel cable along which the motor-truck must haul itself in its progress. This cable will be kept stationary, but gripped by the wheels and other appliances of the electric motors with which the long trucks are provided. Besides this there must also be the conducting cables for the conveyance of the electric current.

For cheap means of transport in sparsely-developed country, as well as in regions of an exceptionally hilly contour, the "wheelway" has a great future before it. Ultimately the system can be worked out so as to present an almost exact converse of the railway. The rails are fixed on the lower part of the elongated truck, one on each side; while the wheels, placed at intervals upon suitable supports, constitute the permanent way. The amount of constructional work required for each mile of track under this plan is a mere fraction of that which is needed for the permanent way and rolling stock of a railway, the almost entire absence of earth-works being, of course, a most important source of economy.

Probably the development of transport on the principles indicated by the evolution of the ropeway or wire-rope tramway will take place primarily in connection with mining properties, and for general transport purposes in country of a nature which renders it unsuitable for railway construction. This applies not merely to hilly regions, but particularly to those long stretches of sandy country which impede the transport of traffic in many rich mining regions, and in patches separating good country from the seaboard. In the "wheelway" for land of this character the wheels need not be elevated more than a very few feet above the ground, just enough to keep them clear of the drift sand which in some places is fatal to the carrying out of any ordinary railway project.

The conception of a truck or other vehicle that shall practically carry its own rail-road has been an attractive one to some inventive minds. In sandy regions, and in other places where a railway track is difficult to maintain, and where, at any rate, there would hardly be sufficient traffic to encourage expenditure in laying an iron road, a very great boon would be a kind of motor which would lay its own rails in front of its wheels and pick them up again as soon as they had passed.

A carriage of this kind was worked for some time on the Landes in France. The track was virtually a kind of endless band which ran round the four wheels, bearing a close resemblance to the ramp upon which the horse is made to tread in the "box" type of horse-gear. Several somewhat similar devices have been brought out, and a gradual approach seems to have been made towards a serviceable vehicle.

A large wheel offers less resistance to the traction of the weight upon it than a small one. The principal reason for this is that its outer periphery, being at any particular point comparatively straight, does not dip down into every hollow of the road, but strikes an average of the depressions and prominences which it meets. The pneumatic tyre accomplishes the same object, although in a different way, the weight being supported by an elastic surface which fits into the contour of the ground beneath it; and the downward pressure being balanced by the sum total of all the resistant forces offered by every part of the tyre which touches the ground, whether resting on hollows or on prominences.

Careful tests which have been made with pneumatic-tyred vehicles by means of various types of dynamometer have proved that, altogether apart from the question of comfort arising from absence of vibration, there is a very true and real saving of actual power in the driving of a vehicle on wheels fitted with inflated tubes, as compared with the quantity that is required to propel the same vehicle when resting on wheels having hard unyielding rims. So far as cycles and motor-cars are concerned, this is the best solution of the problem of averaging the inequalities of a road that has yet been presented; but when we come to consider the making of provision for goods traffic carried by traction engines along ordinary roadways, the difficulties which present themselves militating against the adoption of the pneumatic principle—at any rate so long as a cheap substitute for india-rubber is undiscovered—are practically insurmountable.

Large cart wheels of the ordinary type are much more difficult to construct than small ones, besides being more liable to get out of order. The advantages of a large over a small wheel in reducing the amount of resistance offered by rough roads have long been recognised, and the limit of height was soon attained. In looking for improvement in this direction, therefore, we must inquire what new types of wheel may be suggested, and whether an intermediate plan between the endless band, as already referred to, and the old-fashioned large wheel may not find a useful place.

Let the wheel consist of a very small truck-wheel running on the inside of a large, rigid steel hoop. The latter must be supported, to keep it from falling to either side, by means of a steel semi-circular framework rising from the sides of the vehicle and carrying small wheels to prevent friction. We now have a kind of rail which conforms to the condition already mentioned, namely, that of being capable of being laid down in front of the wheel of the truck or vehicle, and of being picked up again when the weight has passed over any particular part. The hoop, in fact, constitutes a rolling railway, and the larger it can with convenience be made, the nearer is the approach which it presents to a straight railway track as regards the absence of resistance to the passing of a loaded truck-wheel over it.

The method of applying the rolling hoop, more particularly as regards the question whether two or four shall be used for a vehicle, will depend upon the special work to be performed. Some vehicles, however, will have only two hoops, one on each side, but several small truck-wheels running on the inside of each. A vehicle of this pattern is not to be classed with a two-wheeled buggy, because it will maintain its equilibrium without being held in position by shafts or other similar means. So far as contact with the road is concerned it is two-wheeled; and yet, in its relation to the force of gravitation upon which its statical stability depends, it is a four or six-wheeler according to the number of the small truck-wheels with which it is fitted.

Traction engines carrying hoops twenty feet in height, or at any rate as high as may be found compatible with stability when referred to the available width on the road, will be capable of transporting goods at a cost much below that of horse traction. The limit of available height may be increased by the bringing of the two hoops closer to each other at the top than they are at the roadway, because the application of the principle does not demand that the hoops should stand absolutely erect.

Similar means will, no doubt, be tried for the achievement of a modified form of what has been dreamt of by cyclists under the name of a unicycle. This machine will resemble a bicycle running on the inner rim of a hoop, and will probably attain to a higher speed for show purposes than the safety high-geared bicycle of the usual pattern. But it is in the development of goods traffic along ordinary roads that the hoop-rail principle will make its most noticeable progress. By its agency not only will the transport of goods along well-made roads become less costly and more expeditious, but localities in sparsely settled countries—such as those beyond the Missouri in America and the interior regions of South Africa, Australia and China—will become much more readily accessible.

A traction-engine and automobile which can run across broad, almost trackless plains at the rate of fifteen miles an hour will bring within quick reach of civilisation many localities in which at present, for lack of such communication, rough men are apt to grow into semi-savages, while those who retain the instincts of civilisation look upon their exile as a living death. It will do more to enlighten the dark places of the earth than any other mechanical agency of the twentieth century.

CHAPTER VI
SHIPS

The "cargo slave" and the "ocean greyhound" are already differentiated by marked characteristics, and in the twentieth century the divergence between the two types of vessels will become much accentuated. The object aimed at by the owners of cargo boats will be to secure the greatest possible economy of working, combined with a moderately good rate of speed, such as may ensure shippers against having to stand out of their capital locked up in the cargo for too long a period. Hence cheap power will become increasingly a desideratum, and the possible applications of natural sources of energy will be keenly scrutinised with a view to turning any feasible plan to advantage. The sailing ship, and the economic and constructive lines upon which it is built and worked, will be carefully overhauled with a view to finding how its deficiencies may be supplemented and its good points turned to account. One result of this renewed attention will be to confirm, for some little time, the movement which showed itself during the past decade of the nineteenth century for an increase of sailing tonnage. Sooner or later, however, it will be recognised that sail power must be largely supplemented, even on the "sailer," if it is to hold its own against steam.

For mails and passengers, on the other hand, steam must more and more decidedly assert its supremacy. Yet the mail-packet of the twentieth century will be very different from packets which have "made the running" towards the close of the nineteenth. She will carry little or no cargo excepting specie, and goods of exceptionally high value in proportion to their weight and bulk. Nearly all her below-deck capacity, indeed, will be filled with machinery and fuel. She will be in other respects more like a floating hotel than the old ideal of a ship, her cellars, so to speak, being crammed with coal and her upper stories fitted luxuriously for sitting and bed rooms and brilliant with the electric light. But in size she will not necessarily be any larger than the nineteenth century type of mail steamer. Indeed the probability is that, on the average, the twentieth century mail-packets will be smaller, being built for speed rather than for magnificence or carrying capacity.

The turbine-engine will be the main factor in working the approaching revolution in mail steamer construction. The special reason for this will consist in the fact that only by its adoption can the conditions mentioned above be fulfilled. With the ordinary reciprocating type of marine steam machinery it would be impossible to place, in a steamer of moderate tonnage, engines of a size suitable to enable it to attain a very high rate of speed, because the strain and vibration of the gigantic steel arms, pulling and pushing the huge cranks to turn the shafting, would knock the hull to pieces in a very short time. For this very reason, in fact, the marine architect and engineer have hitherto urged, with considerable force of argument, that high speed and large tonnage must go concomitantly. Practically, only a big steamer, with the old type of marine-engine, could be a very fast one, and, for ocean traffic at any rate, a smaller vessel must be regarded as out of the running. Very large tonnage being thus made a prime necessity, it followed that the space provided must be utilised, and this need has tended to perpetuate the combination of mail and passenger traffic with cargo carrying.

The first step towards the revolution was taken many years ago when the screw propeller was substituted for the paddle-wheel. The latter means of propulsion caused shock and vibration not only owing to the thrusts of the piston-rod from the steam-engine itself, but also from the impact of the paddles upon the water one after the other. A great increase in the smoothness of running was attained when the screw was invented—a propeller which was entirely sunk in the water and therefore exercised its force, not in shocks, but in gentle constant pressure upon the fluid around it. Such as the windmill is for wind and the turbine water-wheel for water was the screw propeller, although adapted, not as a generator, but as an application of power. Having made the work and stress continuous, the next thing to be accomplished was to effect a similar reform in the engines supplying the power. This is accomplished in the turbine steam-engine by causing the steam to play in strong jets continuously and steadily upon vanes which form virtually a number of small windmills. Thus, while the screw outside of the hull is applying the force continuously, the steam in the inside is driving the shafting with equal evenness and regularity.

The steam turbine does not appear to have by any means reached finality in its form, such questions as the angle of impact which the jet should make with the surface of the vane, and the size of the orifice through which the steam should be ejected, being still debatable points. But on one matter there is hardly any room for doubt, and that is that the best way to secure the benefit of the expansive power of steam is to permit it to escape from a pipe having a long series of orifices and to impinge upon a correspondingly numerous series of vanes, or, perhaps, upon a number of vanes arranged so that each one is long enough to receive the impact of many jets.

Hitherto the steam supply-pipe emitting the jet has been placed outside of the circle of the wheel; but the future form seems likely to be one in which the axis of the wheel is itself the pipe which contains the steam, but which permits it to escape outwards to the circumference of the wheel. The latter is, in this form of turbine, made in the shape of a paddle-wheel of very small circumference but considerable length, the paddles being set at such an inclination as to obtain the greatest possible rotative impulse from the outward-rushing steam. The pipe must be turned true at intervals to enable it to carry a number of diminutive wheels upon which these long vanes are mounted, and a very strong connection must be made between these wheels and the shaft of the screw. Inasmuch as a high speed of rotation is to be maintained, the pitch of the screw in the water is set so as to offer but slight opposition to the water at each turn. The immense speed attained is thus due, not to the actual power with which the water is struck by the screw at each revolution, but to the extraordinary rapidity with which the shaft rotates.

The twin screw, with which the best and safest of modern steam-ships are all fitted, will soon develop into what may be called "the twin stern". Each screw requires a separate set of engines and the main object of the duplication is to lessen the risk of the vessel being left helpless in case of accident to one or other. The advisability of placing each engine and shafting in a separate water-tight compartment has therefore been seen. At this point there presents itself for consideration the advisability of separating the two screws by as wide a distance as may be convenient and placing the rudder between the two. Practically, therefore, it will be found best to build out a steel framework from each side of the stern for holding the bearings of each screw in connection with the twin water-tight compartments holding the shafting; and thus will be evolved what will practically represent a twin, or double, stern.

In the case of the turbine steamer several of the forms of screw which were first proposed when that type of propeller was invented will again come up for examination, notably the Archimedean screw, wound round a fairly long piece of shafting. The larger the circular area of this screw is the less will be the risk of "smashing" the water, or of losing hold of it entirely in rough weather. With twin screws of the large Archimedean type the propelling apparatus of a turbine steamer will—if the screws are left open— be objected to on the ground of liability to foul or to get broken in crowded fairways. Hence will arise a demand for accommodation for each screw in a tube forming part of the lower hull itself and open at the side for the taking in of water, while the stern part is equally free. In this way there is evolved a kind of compromise between the two principles of marine propulsion, by a screw and by a jet of water thrown to sternward. The water-jet is already very successfully employed for the propulsion of steam lifeboats in which, owing to the danger of fouling the life-saving and other tackle, an open screw is objectionable.

The final extermination of the sailing ship is popularly expected as one of the first developments of the twentieth century in maritime traffic. Steam, which for oversea trade made its entrance cautiously in the shape of a mere auxiliary to sail power, had taken up a much more self-assertive position long before the close of the nineteenth century, and has driven its former ally almost out of the field in large departments of the shipping industry. Yet a curious and interesting counter movement is now taking place on the Pacific Coast of America, as well as among the South Sea Islands and in several other places where coal is exceptionally dear. Trading schooners and barques used in these localities are often fitted with petroleum oil engines, which enable them to continue their voyages during calm or adverse weather. For the owners of the smaller grade of craft it was a material point in recommendation of this movement that, having no boiler or other parts liable to explode and wreck the vessel, an oil engine may be worked without the attendance of a certificated engineer. As soon as this legal question was settled a considerable impetus was given to the extension of the auxiliary principle for sailing ships. The shorter duration of the average voyage made by the sail-and-oil power vessels had the effect of enabling shippers to realise upon the goods carried more speedily than would have been possible under the old system of sail-power alone.

It is already found that in the matter of economy of working, including interest on cost of vessel and cargo, these oil-auxiliary ships can well hold their own against the ordinary steam cargo slave. Up to a certain point, the policy of relying upon steam entirely, unaided by any natural cheap source of power, has been successful; but the rate of speed which the best types of marine engines impart to this kind of vessel is strictly limited, owing to considerations of the enormous increase of fuel-consumption after passing the twelve or fourteen mile grade. For ocean greyhounds carrying mails and passengers the prime necessity of high speed has to a large extent obliterated any such separating line between waste and economy. It is, however, a mistake to imagine that the cargo steamer of the future will be in any sense a replica of the mail-boat of to-day. The opposition presented by the water to the passage of a vessel increases by leaps and bounds as soon as the rate now adopted by the cargo steamer is passed, and thus presents a natural barrier beyond which it will not be economically feasible to advance much further.

If then we recognise clearly that steam cargo transport across the ocean can only be done remuneratively at about one half the speed now attained by the very fastest mail-boats, we shall soon perceive also that the chances of the auxiliary principle, if wisely introduced, placing the "sailer" on a level with the cargo ship worked by steam alone, are by no means hopeless. A type of vessel which can be trusted to make some ten or twelve knots regularly, and which can also take advantage of the power of the wind whenever it is in its favour, must inevitably possess a material advantage over the steam cargo slave in economy of working, while making almost the same average passages as its rival.

Then, also, the sailless cargo slave, in the keen competition that must arise, will be fitted with such appliances as human ingenuity can in future devise, or has already tentatively suggested, for invoking the aid of natural powers in order to supplement the steam-engine and effect a saving in fuel. One of these will no doubt be the adoption of the heavy pendulum with universal joint movement in a special hold of the vessel so connected with an air-compression plant that its movements may continually work to fill a reservoir of air at a high pressure. The marine engines of the ordinary type will then be adapted to work with compressed air, and the true steam-engine itself will be used for operating an air compressor on the system adopted in mines.

The pendulum apparatus, of course, is really a device for enabling a vessel to derive, from the power of the waves which raise her and roll her, an impetus in the desired direction of her course. Inventions of this description will at first be only very cautiously and partially adopted, because if there is one thing which the master mariner fears more than another it is any heavy moving weight in the hold, the motions of which during a storm might possibly become uncontrollable. When steam was first applied to the propulsion of ships the common argument against it was that any machine worked by steam and having sufficient power to propel a vessel would also develop so much vibration as to pull her to pieces—to say nothing of the risk of having her hull shattered at one fell blow by the explosion of the steam boiler. These undoubtedly are dangers which have to be provided against, and probably the occasional lack of care has been the cause of many an unreported loss, as well as of recorded mishaps from broken tail-shafts and screws, or from explosions far out at sea.

The air-compressing pendulum will no doubt be constructed on such a principle that, whenever there is any danger of its weighty movements getting beyond control or doing any damage to the vessel, its force can be instantly removed at will, and the apparatus can be brought to a standstill by the application of friction brakes and other means. The weight may be made up of comparatively small pigs of iron, which, through the opening of a valve controlled from the deck by the stem of the pendulum, can be let fall out into the hold separately. The swinging framework would then be steadied by the friction brake gripping it gradually.

Auxiliary machinery of this class can only be made use of, as already indicated, to a certain strictly limited extent, owing to the tendency of any swinging weight in a vessel to aggravate the rolling during heavy weather. Some tentative schemes have been put forward for tapping a source of wave-power by providing a vessel with flippers, resting upon the surface of the water outside her hull, and actuating suitable internal machinery with the object of propulsion. A certain amount of encouragement has been given by the performances of small craft fitted in this way; but it is objected by sea-faring men that the behaviour of a large vessel, encumbered with outlying parts moving on the waves independently, would probably be very erratic during a storm and would endanger the safety of the ship itself. No kind of floating appendage, moving independently of the vessel, could exercise any actual force by the uprising of a wave in lifting it without being to some extent sunk in the water; and, accordingly, when the waves were running high there would be imminent risk that heavy volumes of water would get upon the apparatus and prevent the ship from righting itself. Many of the schemes that have been put forward, by patent and otherwise, for the automatic propulsion of ships have entirely failed to commend themselves by reason of their taking little or no account of the behaviour of a ship, fitted with the proposed inventions, during very rough and trying weather.

The swinging pendulum, with connected apparatus for compressing air or, perhaps, for generating the electric current, seems to be the most controllable and therefore the safest of the various types of apparatus which are applicable to the utilisation of wave-power for propulsion. In the construction of connecting machinery by which the movements of a pendulum hanging up from a universal joint may be transmitted to wheels or pistons operating compressors or dynamos, it is necessary to transform all motions passing in any direction through the spherical or bowl-shaped figure traced out by the end of the pendulum in the course of its swinging. This may be effected, for instance, in the case of a pendulum working air-compressors, by mounting the latter on bearings like those of the gun-carriage in a field piece, and having two of them operating one at right angles to the other. The rods which carry the air-compressing pistons are then connected to the end of the pendulum by universal joints, and the parts which have been likened to a gun-carriage are fixed on pivots so as to be able to move horizontally. Air-tight joints in the pipes which lead to the compressed air reservoir are placed in the bearings of this mounting. We thus have the same kind of provision for taking advantage of a universal movement in space as is made in solid geometry by three co-ordinates at right angles to one another for measuring such movements.

Another plan is to have the pendulum swung in a strong steel collar and carrying at its end three or more air-compressing pumps set radially, with the piston-rods thrust outwards by a strong spring on each, but with the ends perfectly free from any attachment, yet fitted with a buffer or wheel. As the pendulum moves it throws one or more of these piston-rod ends into contact with the inner surface of the ring, driving it into the compressing pump. At the top of the pendulum there is a double or universal pipe-joint through which the air under pressure is driven to the reservoir, and by which the apparatus is also hung. This is the simplest, and in some respects the best, form.

A very simple type of the wave-power motor as applied to marine propulsion is based upon an idea taken from the mode of progression adopted by certain crustaceans, namely the possession of the means for drawing in and rapidly ejecting the water. Something of the kind will most probably be made available for assisting in the propulsion of sailing ships which are not furnished with machinery of any type suitable for the driving of a screw. A very much simplified form of the pendulous or rocking weight is applicable in this case. A considerable amount of cargo is stowed away in an inner hull, taking the shape of what is practically a gigantic cradle rocking upon semicircular lines of railway iron laid down in the form of ribs of the ship. To the sides of these large rocking receptacles are connected the rods carrying, at their other ends, the pistons of large force-pumps which draw the water in at one stroke and force it out to sternwards, below the water line, at the other.

In this arrangement it is obvious that only the "roll" and not the "pitch" of the vessel can be utilised as the medium through which to obtain propulsive force. But it is probable that fully eighty per cent. of the movements of a vessel during a long voyage—as indicated, say, by the direction and sweep of its mast-heads—consists of the roll. Each ton of goods moved through a vertical distance of one foot in relation to the hull of the vessel, has in it the potentiality of developing, when fourteen or fifteen movements occur per minute, about one horse-power. A cradle containing 200 tons, as may therefore be imagined, can be made to afford very material assistance in helping forward a sailing ship during a calm. In such tantalising weather the "ground-swell" of the ocean usually carries past a becalmed vessel more waste energy than is ever utilised by its sails in the briskest and most propitious breeze.

For sailing ships especially, the rocking form of wave-motor as an aid to propulsion will be recommended on account of the fact that when the weather is "on the beam" both of its sources of power can be kept in full use. The sailing vessel must tack at any rate with the object of giving its sail power a fair chance, and thus, when it has not a fair "wind that follows free," it must always seek to get the breeze on its beam, and therefore usually the swell must be taking it sideways. It would be only on rare occasions that a sailing vessel, if furnished with rocking gear for using the wave-power, would be set to go nearer to the teeth of the wind than she would under present conditions of using sail-power alone. The advantage of the wave-power, however, would be seen mainly during the calm and desultory weather which has virtually been the means of forcing sail-power to resign its supremacy to steam.

For checking the rocker in time of heavy weather special appliances are necessary, which, of course, must be easily operated from the deck. Wedge-shaped pieces with rails attached may be driven down by screws upon the sides of the vessel, thus having the effect of gradually narrowing the amplitude of the rocking motion until a condition of stability with reference to the hull has been attained.

In the building of steel ships, as well as in the construction of bridges and other erections demanding much metal-work, great economies will be introduced by the reduction of the extent to which riveting will be required when the full advantages of hydraulic pressure are realised. The plates used in the building of a ship will be "knocked-up" at one side and split at the other, with the object of making joints without the need for using rivets to anything like the extent at present required. In putting the plates thus treated together to form the hull of a vessel the swollen side of one plate is inserted between the split portions of another and the latter parts are then clamped down by heavy hydraulic pressure. This important principle is already successfully used in the making of rivetless pipes, and its application to ships and bridges will be only a matter of a comparatively short time. Through this reform, and the further use of steel ribs for imparting strength and thus admitting of the employment of thinner steel plates for the actual shell, the cost of shipbuilding will be very greatly reduced.

Hoisting and unloading machines will play a notable part in minimising the expenses of handling goods carried by sea. The grain-elevator system is only the beginning of a revolution in this department which will not end until the loading and unloading of ships have become almost entirely the work of machinery. The principle of the miner's tool known as the "sand-auger" may prove itself very useful in this connection. From a heap of tailings the miner can select a sample, by boring into it with a thin tube, inside of which revolves a shaft carrying at its end a flat steel rotary scoop. The auger, after working its way to the bottom of the heap, is raised, and, of course, it contains a fair sample of the sand at all depths from the top downwards. On a somewhat similar principle the unloading of ships laden with grain, ore, coal, and all other articles which can be handled in bulk and divided, will be carried out by machines which, by rotary action, will work their way down to the bottom of the hull and will then be elevated by powerful lifting cranes. For other classes of goods permanent packages and tramways will be provided in each ship, and trucks will be supplied at the wharf.

For coastal passages across shallow but rough water like the English Channel, the services of moving bridges will be called into requisition. One of these has been at work at St. Malo on the French coast opposite Jersey, and another was more recently constructed on the English coast near Brighton. For the longer and much more important service across the Channel submarine rails may be laid down as in the cases mentioned, but in addition it will be necessary to provide for static stability by fixing a flounder-shaped pontoon just below the greatest depth of wave disturbance, and just sufficient in buoyancy to take the great bulk of the weight of the structure off the rails. In this way passengers may be conveyed across straits like the Channel without the discomforts of sea-sickness.

The stoking difficulties on large ocean-going steamers have become so acute that they now suggest the conclusion that, notwithstanding repeated failures, a really effective mechanical stoker will be so imperatively called for as to enforce the adoption of any reasonably good device. The heat, grime, and general misery of the stoke-hole have become so deterrent that the difficulty of securing men to undertake the work grows greater year by year, and in recruiting the ranks of the stokers resort had to be had more and more to those unfortunate men whose principal motive for labour is the insatiable desire for a drinking bout. On the occasions of several shipwrecks in the latter part of the nineteenth century disquieting revelations took place showing how savagely bitter was the feeling of the stoke-hole towards the first saloon. As soon as the mechanical fuel-shifter has been adopted, and the boilers have been properly insulated in order to prevent the overheating of the stoke-hole, the stoker will be raised to the rank of a secondary engineer, and his work will cease to be looked upon as in any sense degrading.

On the cargo-slave steamer and sailer a similar social revolution will be brought about by the amelioration of the conditions under which the men live and work. Already some owners and masters have begun to mitigate, to a certain extent, the embargo which the choice of a sea-faring life has in times past been understood to place upon married men. Positions are found for women as stewardesses and in other capacities, and it is coming to be increasingly recognised that there is a large amount of women's work to be done on board a ship.

By and by, when it is found that the best and steadiest men can be secured by making some little concessions to their desire for a settled life and their objections to the crimp and the "girl at every port," and all the other squalid accessories so generally attached in the popular mind to the seaman's career, there will be a serious effort on the part of owners to remodel the community on board of a ship on the lines of a village. There will be the "Ship's Shop" and the "Ship's School," the "Ship's Church" and various other institutions and societies.

Thus in the twentieth century the sea will no longer be regarded, to the same extent as in the past, as the refuge for the ne'er-do-well of the land-living populace; and this, more than perhaps anything else, will help to render travelling by the great ocean highways safe and comfortable. It is a common complaint on the part of owners that by far the larger part of maritime disasters are directly traceable to misconduct or neglect of duty on the part of masters, officers or crew; but they have the remedy in their own hands.

CHAPTER VII
AGRICULTURE

Muscular power still carries out all the most laborious work of the farm and of the garden—work which, of course, consists, in the main, of turning the land over and breaking up the sods. In the operations of ploughing, harrowing, rolling, and so forth, the agency almost exclusively employed is the muscular power of the horse guided by man-power; with the accompaniment of a very large and exhausting expenditure of muscular effort on the part of the farmer or farm labourer. On the fruit and vegetable garden the great preponderance of the power usefully exercised must, under existing conditions, come direct from the muscles of men. Spade and plough represent the badges of the rural workers' servitude, and to rescue the country residents from this old-world bondage must be one of the chief objects to which invention will in the near future apply itself.

The miner has to a very large extent escaped from the thraldom of mere brute-work, or hardening muscular effort. He drills the holes in the face of the rock at which he is working by means of compressed air or power conveyed by the electric current; and then he performs the work of breaking it down by the agency of dynamite or some other high explosive. Much heavy bodily labour, no doubt, remains to be done by some classes of workers in mines; but the significance of the march of improvement is shown by the fact that a larger and larger proportion of those who work under the surface of the ground, or in ore-reduction works, consists of men who are gradually being enrolled among the ranks of the more highly skilled and intelligent workers, whose duty it is to understand and to superintend pieces of mechanism driven by mechanical power.

In farming and horticulture the field of labour is not so narrowly localised as it is in mining. Work representing an expenditure of hundreds of thousands of pounds may be carried out in mines whose area does not exceed two or three acres; and it is therefore highly remunerative to concentrate mechanical power upon such enterprises in the most up-to-date machinery. But the farmer ranges from side to side of his wide fields, covering hundreds, or even thousands, of acres with his operations. He is better situated than the miner in respect of the economical and healthy application of horse-power, but far worse in regard to the immediate possibilities of steam-power and electrically-conducted energy. No one can feed draught stock more cheaply than he, and no one can secure able-bodied men to work from sunrise till evening at a lower wage.

Yet the course of industrial evolution, which has made so much progress in the mine and the factory, must very soon powerfully affect agriculture. Already, in farming districts contiguous to unlimited supplies of cheap power from waterfalls, schemes have been set on foot for the supply of power on co-operative principles to the farmers of fertile land in America, Germany, France, and Great Britain. One necessity which will most materially aid in spurring forward the movement for the distribution of power for rural work is the requirement of special means for lifting water for irrigation, more particularly in those places where good land lies very close to the area that is naturally irrigable, by some scheme already in operation but just a little too high. Here it is seen at once that power means fertility and consequent wealth, while the lack of it—if the climate be really dry, as in the Pacific States of America—means loss and dearth. But the presence of a source of power which can easily be shifted about from place to place on the farm for the purpose of watering the ground must very soon suggest the applicability of the same mechanical energy to the digging or ploughing of the soil.

It is from this direction, rather than from the wide introduction of steam-ploughs and diggers, that the first great impetus to the employment of mechanical power on the farm may be looked for. The steam-plough, no doubt, has before it a future full of usefulness; and yet the slow progress that has been made by it during a quarter of a century suggests that, in its present form—that is to say while built on lines imitating the locomotive and the traction-engine—it cannot very successfully challenge the plough drawn by horse-power. More probable is it—as has already been indicated—that the analogy of the rock-drill in mining work will be followed. The farmer will use an implement much smaller and handier than a movable steam-engine, but supplied with power from a central station, either on his own land or in some place maintained by co-operative or public agency. Just as the miner pounds away at the rock by means of compressed air or electricity, brought to his hands through a pipe or a wire, so the farmer will work his land by spades or ploughs by the same kind of mechanical power. The advantages of electrical transmission of energy will greatly favour this kind of installation on the farm, as compared with any other method of distribution which is as yet in sight.

For the ploughing of a field by the electric plough a cable will be required capable of being stretched along one side of the area to be worked.

On this will run loosely a link or wheel connected with another wire wound upon a drum carried on the plough and paid out as the latter proceeds across the field. For different grades of land, of course, different modes of working are advisable, the ordinary plough of a multifurrow pattern, with stump-jumping springs or weights, being used for land which is not too heavy or clayey; a disc plough or harrow being applicable to light, well-worked ground; and the mechanical spade or fork-digger—reciprocating in its motion very much like the rock-drill—having its special sphere of usefulness in wet and heavy land. In any case a wide, gripping wheel is required in front to carry the machine forward and to turn it on reaching the end of the furrow. The wire-wound drum is actuated by a spring which tends to keep it constantly wound up, and when the plough has turned and is heading again towards the cable at the side of the field, this drum automatically winds up the wire. So also when each pair of furrows has been completed, the supply-wire is automatically shifted along upon the fixed cable to a position suitable for the next pair.

Not only in the working, but also in the manuring, of the soil the electric current will play an important part in the revolution in agriculture. The fixing of the nitrogen from the atmosphere in order to form nitrates available as manure depends, from the physical point of view, upon the creation of a sufficient heat to set fire to it. The economic bearings of this fact upon the future of agriculture, especially in its relation to wheat-growing, seemed so important to Sir William Crookes that he made the subject the principal topic of his Presidential Address before the British Association in 1898.

The feasibility of the electrical mode of fixing atmospheric nitrogen for plant-food has been demonstrated by eminent electricians, the famous Hungarian inventor, Nikola Tesla, being among the foremost. The electric furnace is just as readily applicable for forcing the combination of an intractable element, such as nitrogen, with other materials suitable for forming a manurial base, as it is for making calcium carbide by bringing about the union of two such unsociable constituents as lime and carbon.

Cheap power is, in this view, the great essential for economically enriching the soil, as well as for turning it over and preparing it for the reception of seed. Nor is the fact a matter of slight importance that this power is specially demanded for the production of an electric current for heating purposes, because the transmission of such a current over long distances to the places at which the manurial product is required will save the cost of much transport of heavy material.

The agricultural chemist and the microbiologist of the latter end of the nineteenth century have laid considerable stress upon the prospects of using the minute organisms which attach themselves to the roots of some plants—particularly those of the leguminaceæ—as the means of fixing the nitrogen of the atmosphere, and rendering it available for the plant-food of cereals which are not endowed with the faculty of encouraging those bacteria which fix nitrogen. High hopes have been based upon the prospects of inoculating the soil over wide areas of land with small quantities of sandy loam, taken from patches cultivated for leguminous plants which have been permitted to run to seed, thus multiplying the nitrogen-fixing bacteria enormously. The main idea has been to encourage the rapid production of these minute organisms in places where they may be specially useful, but in which they do not find a particularly congenial breeding ground.

The hope that any striking revolution may be brought about in the practice of agriculture by a device of this kind must be viewed in the light of the fact that, while the scientists of the nineteenth century have demonstrated, partially at least, the true reason for the beneficial effects of growing leguminous plants upon soil intended to be afterwards laid down in cereals, they were not by any means the first to observe the fact that such benefits accrued from the practice indicated. Several references in the writings of ancient Greek and Latin poets prove definitely that the good results of a rotation of crops, regulated by the introduction of leguminous plants at certain stages, were empirically understood. In that more primitive process of reasoning which proceeds upon the assumption *post hoc, ergo propter hoc*, the ancient agriculturist was a past-master, and the chance of gleaning something valuable from the field of common observation over which he has trod is not very great.

Modern improvements in agriculture will probably be, in the main, such as are based upon fundamental processes unknown to the ancients. By the word "processes" it is intended to indicate not those methods the scientific reasons for which were understood—for these in ancient times were very few—but simply those which from long experience were noticed to be beneficial. Good husbandry was in olden times clearly understood to include the practice of the rotation of crops, and the beneficial results to be expected from the introduction of those crops which are now discovered to act as hosts to the microbes which fix atmospheric nitrogen were not only observed, but insisted upon.

From a scientific point of view this concurrence of the results of ancient and of modern observation may only serve to render the bacteriology of the soil more interesting; but, from the standpoint of an estimate of the practical openings for agriculture improvements in the near future, it greatly dwarfs the prospect of any epoch-making change actually founded upon the principle of the rotation of crops. It is, indeed, conceivable that fresh light on the life habits of the minute organisms of the soil may lead to practical results quite new; but hardly any such light is yet within the inventor's field of vision.

This view of the limited prospects of practical microbiology for the fixing of nitrogen in plant-food was corroborated by Sir William Crookes in the Presidential Address already cited. He said that "practice has for a very long time been ahead of science in respect of this department of husbandry". For ages what is known as the four course rotation had been practised, the crops following one another in this order—turnips, barley, clover and wheat—a sequence which was popular more than two thousand years ago. His summing up of the position was to the effect that "our present knowledge leads to the conclusion that the much more frequent growth of clover on the same land, even with successful microbe-seeding and proper mineral supplies, would be attended with uncertainties and difficulties, because the land soon becomes what is called clover-sick, and turns barren".

In regard to any practical application of microbe-seeding, the farmers of the United Kingdom at the end of the nineteenth century had not, in the opinion of this eminent chemist, reached even the experimental stage, although on the Continent there had been some extension of microbe cultivation. To this it may fairly be added that some of the attention attracted to the subject on the Continent has been due to the natural tendency of the German mind to discover fine differences between things which are not radically distinct. Under the title of "microbe-cultivation" the long-familiar practice of the rotation of crops may to some continental enthusiasts seem to be quite an innovation!

In the electrical manures-factory the operations will be simply an enlargement of laboratory experiments which have been familiar to the chemist for many years. Moist air, kept damp by steam, is traversed by strong electric sparks from an induction coil inside of a bottle or other liquor-tight receiver, and in a short time it is found that in the bottom of this receptacle a liquid has accumulated which, on being tested, proves to be nitric acid. There is also present a small quantity of ammonia from the atmosphere. Nitrate of ammonia thus formed would in itself be a manure; but, of course, on the large scale other nitrates will be formed by mixing the acid with cheap alkalies which are abundant in nature, soda from common salt, and lime from limestone.

In this process the excessive heat of the electric discharge really raises the nitrogen and oxygen of the atmosphere to a point of temperature at which chemical union is forced; or, in other words, the nitrogen is compelled to burn and to join in chemical combination with the oxygen with which formerly it was only in mechanical mixture. When nitrogen is burning, its flame is not in itself hot enough to ignite contiguous volumes of the same element;—otherwise indeed our atmosphere, after a discharge of lightning, would burn itself out!—but the continuance of an electric discharge forces into combination just a proportionate quantity of nitrogen. Practically, therefore, manure in the future will mean electricity, and therefore power; so that cheap sources of energy are of the greatest importance to the farmer.

With dynamos driven by steam-engines, the price of electrically-manufactured nitrate of soda would, according to the estimate of Sir William Crookes, be £26 per ton, but at Niagara, where water power is very cheap, not more than £5 per ton. Thus it will be seen that the cheapness of power due to the presence of the waterfall makes such a difference in the economic aspects of the problem of the electrical manufacture of manurial nitrates as to reduce the price to less than one-fifth! It must be remembered that at the close of the nineteenth century the electric installation at Niagara is by very many persons looked upon as being in itself in the nature of an experiment, but at any rate there seems to be no room for doubt that the cost of natural power for electrical installations will very soon be materially reduced. Even at the price quoted, namely £5 per ton, the cost of nitrate of soda made with electrically combined atmospheric nitrogen compares very favourably with commercial nitrates as now imported for agriculture purposes. "Chili nitrate," in fact, is about fifty per cent. dearer.

When wave-power and other forms of the stored energy of the wind have been properly harnessed in the service of mankind, the region around Niagara will only be one of thousands of localities at which nitrogenous manures can be manufactured electrically at a price far below the present cost of natural deposits of nitrate of soda. From the power stations all around the coasts, as well as from those on waterfalls and windy heights among the mountains, electric cables will be employed to convey the current for fixing the nitrogen of the air at places where the manures are most wanted.

The rediscovery of the art of irrigation is one of the distinguishing features of modern industrial progress in agriculture. Extensive ruins and other remains in Assyria, Egypt, India, China and Central America prove beyond question that irrigation played a vastly more important part in the industrial life of the ancients than it does in that of modern mankind. This is true in spite of the fact that power and dominion ultimately fell to the lot of those races which originally dwelt in colder and more hilly or thickly-wooded regions, where the instincts of hunting and of warfare were naturally developed, so that, by degrees, the peoples who understood irrigation fell under the sway of those who neither needed nor appreciated it. In the long interval vast forests have been cleared away and the warlike habits of the northern and mountainous races have been greatly modified, but manufacturing progress among them has enabled them to perpetuate the power originally secured by the bow and the spear. The irrigating races of mankind are now held in fear of the modern weapons which are the products of the iron and steel industries, just as they were thousands of years ago terrorised by the inroads of the wild hunting men from the North.

But the future of agriculture will very largely belong to a class of men who will combine in themselves the best attributes of the irrigationist and the man who knows how to use the iron weapon and the iron implement. As the manufacturing supremacy of the North becomes more and more assured by reason of the superior healthiness of a climate encouraging activity of muscle and brain, so the agricultural prospects of the warmer regions of the earth's surface will be improved by the comparative immunity of plant and of animal life from disease in a dry atmosphere. Sheep, cattle and horses thrive far better in a climate having but a scanty rainfall than in one having an abundance of wet; and so, also, does the wheat plant when the limited rains happen to be timed to suit its growth, and the best kinds of fruit trees when the same conditions prevail.

All this points to an immense recrudescence of irrigation in the near future. Already the Californians and other Americans of the Pacific Slope have demonstrated that irrigation is a practice fully as well suited to the requirements of a thoroughly up-to-date people as it has been for long ages to those of the "unchanging East". But here again the question of cheap power obtrudes itself. The Chinese, Hindoos and Egyptians have long ago passed the stage at which the limited areas which were irrigable by gravitation, without advanced methods of engineering, have been occupied; and the lifting of water for the supplying of their paddy fields has been for thousands of years a laborious occupation for the poorest and most degraded of the rural population.

In a system of civilisation in which transport costs so little as it does in railway and steam-ship freights, the patches of territory which can be irrigated by water brought by gravitation from the hills or from the upper reaches of rivers are comparatively easy of access to a market. This fact retards the advent of the time when colossal installations for the throwing of water upon the land will be demanded. When that epoch arrives, as it assuredly will before the first half of the twentieth century has been nearly past, the pumping plants devoted to the purposes of irrigation will present as great a contrast to the lifting appliances of the East as does a fully loaded freight train or a mammoth steam cargo-slave to a coolie carrier.

At the same time there must inevitably be a great extension of the useful purposes to which small motors can be applied in irrigation. Year by year the importance of the sprinkler, not only for ornamental grounds such as lawns and flower-beds, but also for the vegetable patch and the fruit garden, becomes more apparent, and efforts are being made towards the enlargement of the arms of sprinkling contrivances to such an extent as to enable them to throw a fine shower of water over a very large area of ground. Sometimes a windmill is used for pumping river or well-water into high tanks from which it descends by gravitation into the sprinklers, the latter being operated by the power of the liquid as it descends. This mode of working is convenient in many cases; but a more important, because a more widely applicable, method in the future will be that in which the wind-motor not only lifts the water, but scatters it around in the same operation. Long helical-shaped screws, horizontally fixed between uprights or set on a swivel on a single high tower, can be used for loading the breeze with a finely divided shower of water and thus projecting the moisture to very long distances. A windmill of the ordinary pattern, as used for gardens, may be fitted with a long perforated pipe, supported by wire guys instead of a vane, a connection being made by a water-tight swivel-joint between this pipe and that which carries the liquid from the pump. In this way every stroke of the machine sends innumerable jets of water out upon the wind, to be carried far afield.

Gardening properties in comparatively dry climates, fitted with machines of this description, can be laid out in different zones of cultivation, determined according to the prevailing directions of the wind and the consequent distribution of the water supply. Thus if the wind most frequently blows from the west the plants which require the most water must be laid out at the eastern side, not too far from the sprinkler. Facilities for shutting off the supply of spray at will are, of course, very necessary. The system of watering founded on this principle depends upon the assumption that if the gardener or the farmer could always turn on the rain when he has a fairly good wind he would never lack for seasonable moisture to nourish his crops. This will be found in practice to apply correctly to the great majority of food plants. In the dry climates, which are so eminently healthy for cereals, "the early and the latter rains," as referred to in Scripture, are both needed, and one of the most important applications of cheap power will be directed to supplementing the natural supply either at one end or at the other.

The "tree-doctor" will be a personage of increasing importance in the rural economy of the twentieth century. He is already well in sight; but for lack of capital and of a due appreciation of the value of his services, he occupies as yet but a comparatively subordinate position. Fruits, which are nature's most elaborately worked-up edible products, must come more and more into favour as the complement to the seed food represented by bread. As the demand increases it will be more clearly seen that an enormous waste of labour is involved in the culture of an orchard unless its trees are kept in perfect health. At the same time the law of specialization must operate to set aside the tree-doctor to his separate duties, just as the physician and the veterinary surgeon already find their own distinctive spheres of work. The apparatus required for the thorough eradication of disease in fruit trees will be too expensive for the average grower to find any advantage in buying it for use only a few times during the year; but the tree-doctor, with his gangs of men, will be able to keep his special appliances at work nearly all the year round.

For the destruction of almost all classes of fruit-pests, the only really complete method now in sight is the application of a poisonous gas, such as hydrocyanic acid, which is retained by means of a gas-proof tent pitched around each tree. No kind of a spray or wash can penetrate between bark and stem or into the cavities on fruit so well as a gaseous insecticide which permeates the whole of the air within the included space. But the gas-tight tent system of fumigation is as yet only in its infancy, and its growth and development will greatly help to place the fruit-growing industry on a new basis, and to bring the best kinds of fruit within the reach of the middle classes, the artisans, and ultimately even the very poor. Just as wheaten bread from being a luxury reserved for the rich has become the staple of food for all grades of society, so fruits which are now commonly regarded as an indulgence, although a very desirable addition to the food of the well-to-do, must, in a short time, become practically a necessity to the great mass of the people generally.

The waste of effort and of wealth involved in planting trees and assiduously cultivating the soil for the growth of poor crops decimated by disease is the prime cause of the dearness of fruit. If, therefore, it be true that the fruit diet is one which is destined to greatly improve the average health of civilised mankind, it is obvious that the tree-doctor will act indirectly as the physician for human ailments. When this fact has been fully realised the public estimation in which economic entomology and kindred sciences are held will rise very appreciably, and the capital invested in complete apparatus for fighting disease in tree life will be enormously increased.

Very long tents, capable of covering not merely one tree each, but of including continuous rows stretching perhaps from end to end of a large orchard, will become practically essential for up-to-date fruit-culture. An elongated tent of this description, covering a row of trees, may be filled with fumes from a position at the end of the row, where a generating plant on a trolley may be situated. At the opposite end another trolley is stationed, and each movable vehicle carries an upright mast or trestle for the support of the strong cable which passes along the row over the tops of the trees and is stretched taut by suitable contrivances. Attached to this cable is a flexible tube containing a number of apertures and connected at the generating station with the small furnace or fumigating box from which the poisonous gases emanate.

Along the ground at each side of the row are stretched two thinner wires or cables which hold the long tent securely in position. The method of shifting from one row to another is very simple. Both trolleys are moved into their new positions at the two ends of a fresh row, the fastenings of the tent at the ground on the further side having been released, so that the flap of the tent on that side is dragged over the tops of the trees and may then be drawn over the top cable and down upon the other side. Seen from the end, the movements of the tent thus resemble those of a double-hinged trestle in the form of an inverted V which advances by having one leg flung over the other. For this arrangement of a fumigating tent it is best that the top cable should consist of a double wire, the fabric of the tent itself being gripped between the two wires, and a flexible tube being attached to each.

As progress is made from one row to another through the drawing of one flap over the other, it is obvious that the tent turns inside out at each step, and if only one cable and one tube were used, it would be difficult to avoid permitting the gas to escape into the outer air at one stage or another. But when the tubes are duplicated in the manner described, there is always one which is actually within the tent no matter what position the latter may be in. It is then only necessary that the connection with the generating apparatus at the end of the row should be made after each movement with the tube which is inside the tent. For very long rows of trees the top cable needs to be supported by intermediate trestles besides the uprights at the ends.

The gas and air-proof tent can be used for various other purposes besides those of killing pests on fruit trees. One of the regular tasks of the tree-doctor will be connected with the artificial fertilisation of trees on the wholesale scale and for a purpose such as this it is necessary that the trees to be operated upon shall not be open to the outside atmosphere, but that the pollen dust, with which the air inside the tent is to be laden, shall be strictly confined during a stated period of time. Those methods of fertilisation, with which the flower-gardener has in recent years worked such wonders, can undoubtedly be utilised for many objects besides those of the variation of form and hue in ornamental plants.

CHAPTER VIII
MINING

Exploratory telegraphy seems likely to claim a position in the twentieth century economics of mining, its particular rôle being to aid in the determination of the "strike" of mineral-bearing lodes. One main reason for this conclusion consists in the fact that the formations which carry metalliferous ores are nearly always more moist than the surrounding country, and are therefore better conductors of the electrical current. Indeed there is good ground for the belief that this moistness of the fissures and lodes in which metals chiefly occur has been in part the original cause of the deposition of those metals from their aqueous solutions percolating along the routes in which gravitation carries them. In the volumes of *Nature* for 1890 and 1891 will be found communications in which the present writer has set forth some of the arguments tending to strengthen the hypothesis that earth-currents of electricity exercise an appreciable influence in determining the occurrence of gold and silver, and that they have probably been to some extent instrumental in settling the distribution of other metals.

The existence of currents of electricity passing through the earth's crust and on its surface along the lines of least resistance has long been an established fact. Experiments conducted at Harvard, U.S.A., by Professor Trowbridge have proved beyond a doubt that, by means of such delicate apparatus as the telephone and microphone, it is possible for the observer to state in which direction, from a given point, the best line of conductivity runs. Under certain conditions the return current is so materially facilitated when brought along the line of a watercourse or a moist patch of the earth's crust, that the words heard through a telephone are distinctly more audible than they are at a similar distance when there is no moist return circuit. Deflections of the compass, due to the passing of earth-currents along the natural lines of conductivity in the soil or the rocks, are so frequently noticed as to be a source of calculation to the scientific surveyor and astronomer. It can thus be shown not only that definite lines of least electrical resistance exist in the earth, but also that natural currents of greater or less strength are almost constantly passing along these lines.

Some of the curious and puzzling empirical rules gained from the life-long experience of miners in regard to the varying richness and poorness of mineral lodes, according to the directions in which they strike—whether north, south, east or west—may very probably be explained, and to some extent justified, by the fuller light which science may throw upon the conditions determining the action of earth-currents in producing results similar to those of electro deposition. If, in a given region of a mineral-bearing country, the geological formation is such as to lend itself to the easy conduction of currents in one direction rather than in another, the phenomenon referred to may perhaps be partially explained. But, on the other hand, the origin of the generating force which sets the currents in motion must first be studied before the true conditions determining their direction can be understood. In other words, much that is now obscure, including the true origin of the earth's magnetism, must be to some extent cleared up before the reasons for the seemingly erratic strike of earth-currents and of richness in mineral lodes can be fully explained.

Practice, however, may here get some distance ahead of science, and may indeed lend some assistance to the latter by providing empirical data upon which it may proceed. When once it is clearly seen that by delicate electrical instruments, such as the telephone, the microphone and the coherer as used in wireless telegraphy, the line of least resistance on any given area of the earth's surface or any given piece of its crust may be determined, the bearing of that fact in showing the best lines of moisture and therefore the likeliest lines for mineral lodes will soon be recognised in a very practical manner.

No class of men is keener or more enterprising in its applications of the latest practical science to the getting of money than mining speculators. Nor have they at all missed the significance of moist bands occurring in any underground workings as a very favourable augury for the close approach of highly mineralised lodes. If, then, moisture be favourable, first to the presence of mineral-bearing country and secondly to the conductivity of electrical lines, it is obvious that there is a hopeful field for the exercise of ingenuity in bringing the one into a practical relation to the other.

The occult scientific reasons for the connection may not be understood; but it is sufficient for practical purposes to know that, in a certain line from the surface outcropping of a mineral lode, there has been given a demonstration of less electrical resistance along that line than is experienced in any other direction; also to know that such a line of least resistance is proved to have been, in almost innumerable instances, coincident with the best line of mineral-bearing country. The case is similar to that of the rotation of crops in its relation to scientific microbiology. The art of mining may get ahead of the science of physiography in respect of earth-currents and lines of least resistance, as showing where mineral lodes may be expected. Yet there is no doubt whatever that science will not in the one case lag so far behind as it has done in the other.

The first notable service rendered by systems of the kind indicated will no doubt be in connection with the rediscovery of very valuable lodes which have been followed up for certain distances and then lost. In an instance of this description much fruitless exploration drives, winzes and "jump-ups" may have been carried out in the surrounding country rock near the place where the lode last "cut out"; but, in the absence of anything to guide the mine manager and surveyor as to the direction which the search should take, nothing but loss has been involved in the quest. Several properties in the same neighbourhood have, perhaps, been abandoned or suspended in operation owing to very similar causes.

The whole group may perhaps have then been bought by an exploration company whose *modus operandi* will be as follows: The terminal of the electrical exploration plant is fixed at the end of the lode where it gave out, or else immersed in the water of the shaft which is in connection with the lode system; and another similar terminal is fixed by turns in each shaft of the contiguous group. The electrical resistances offered to the return currents, or to the wireless vibrations, are then carefully measured; and the direction of the lost lode is taken to be that which shows the least resistance in proportion to the distance traversed. The work of carrying out such an investigation must of necessity be somewhat elaborate, because it may be necessary to connect in turn each shaft, as a centre, with every one of the others as subsidiaries. But the guidance afforded even of a negative character, resulting in the avoidance of useless cutting and blasting through heavy country, will prove invaluable.

Many matters will require attention, in following out such a line of practical investigation, which are to some extent foreign to the usual work of the mining engineer. For example, the conditions which determine the "short-circuiting" of an earth-current require to be carefully noted, because it would be fallacious to reason that because the line of least resistance lay in a certain direction, therefore an almost continuous lode would be found. Moreover, the electrical method must only be relied upon as a guide when carefully checked by other considerations. Other kinds of moist formations, both metalliferous and non-metalliferous, may influence the lines of least electrical resistance, besides those containing the particular metal which is being sought for.

The water difficulty has enforced the abandonment of very many valuable mines in which the positions of the lodes are still well known. Sunken riches lying beneath the sea in old Spanish galleons have excited the cupidity and the ingenuity of speculators and engineers; but the total amount of wealth thus hidden away from view is a mere insignificant fraction of the value of the rich metalliferous lodes which lie below the water level in flooded mines.

The point in depth at which the accumulation of the water renders further following of the lode impracticable may vary in different countries. In China, throughout whole provinces, there is hardly a mine to be found in which the efforts of the miners have not been absolutely paralyzed directly the water-level was reached. But in Western lands, as well as in South Africa and Australia, the immense capacity of the pumps employed for keeping down the water has enabled comparatively wet ground to be worked to a very considerable depth.

The limit, nevertheless, has been reached in many rich mining districts. Pumps of the most approved type, and driven by the largest and most economical steam-engines, have done their best in the struggle against the difficulty; and yet the water has beaten them. Rich as are the lodes which lie beneath the water, the mining engineer is compelled to confess that the metal value which they contain would not leave, after extraction, a sufficient margin to pay for the enormous cost of draining the shafts. In some instances, indeed, it remains exceedingly doubtful whether pumps of the largest capacity ever attained in any part of the world would cope with the task entailed in draining the abandoned shafts. The underground workings have practically tapped subterranean rivers which, to all intents and purposes, are inexhaustible. Or it may be that the mine has penetrated into some hollow basin of impermeable strata filled only with porous material which is kept constantly saturated. To drain such a piece of country would mean practically the emptying of a lake.

Subaqueous mining is therefore one of the big problems which the mining engineer of the twentieth century must tackle. To a certain extent he will receive guidance in his difficult task from the experiences of those who have virtually undertaken submarine mining when in search of treasure lost in sunken ships. The two methods of pumping and of subaqueous mining will in some places be carried out conjointly.

In such instances the work assigned to the pumping machinery will be to keep free of water those drives in which good bodies of ore were exposed when last profitable work was being carried on. All below that level will be permitted to fill with water, and the work of boring by means of compressed air, of blasting out the rock and of filling the trucks, will all be performed under the surface. For the shallower depths large tanks, open at the top, will be constructed and slung upon trucks run on rails along the lowest drives. Practically this arrangement means that an iron shaft, closed at the sides and bottom, and movable on rails laid above the surface, will be employed to keep the water out. Somewhat similar appliances have been found very useful in the operations for laying the foundations of bridges.

The details requiring to be worked out for the successful working of subaqueous systems of mining are numerous and important. Chief among these must be the needful provision for enabling the miner to see through strong glass windows near the bottom of the iron shaft, by the aid of electric lights slung in the water outside, and thus to estimate the correct positions at which to place his drills and his explosives. For this reason the work of the day must be systematically divided so that at stated intervals the clay and other materials held in suspension by the disturbed water may be allowed to settle and the water be made comparatively clear.

Specially constructed strainers for the mechanical filtration of the water near the ore face, and probably, also, chemical and other precipitates, will be largely resorted to for facilitating this important operation. Beside each window will be provided strong flexible sleeves, terminating in gloves into which the miner can place his hands for the purpose of adjusting the various pieces of machinery required. Beyond this, of course, every possible application of mechanical power operated from above will be resorted to, not only for drilling, but also for gripping and removing the shattered pieces of rock and ore resulting from the blasting operations.

From the unwatered drive or tunnel downwards, the method of working as just described may be characterised as an underground application of the "open-cut system". No elaborate honeycombing of the country below the water-level will be economically possible as it is when working in dry rock. But then, again, it is becoming plain to many experts in mining that, in working downwards from the surface itself, the future of their industry offers a wide field for the extension of the open-cut system. In proportion as power becomes cheaper, the expense attendant upon the removal of clay, sand, and rock for the purpose of laying bare the cap of a lode at a moderate depth becomes less formidable when balanced against the economy introduced by methods which admit of the miner working in the open air, although at the bottom of a kind of deep quarry. While the system of close mining will hold its own in a very large number of localities, still there are other places where the increasing cheapness of power for working an open-cut and the coincident increase in the scarcity and cost of timber for supporting the ground, will gradually shift the balance of advantage on to the side of the open method.

At the same time great improvements are now foreshadowed in regard to the modes of working mines by shafts and drives. Some shafts will in future be worked practically as the vertical portions of tramways, having endless wire ropes to convey the trucks direct from the face or the stope to the reduction works, and thus an immense saving will be effected in the costs incidental to mining. From the neighbourhood of the place at which it has been won, the ore will be drawn in trucks, attached to the endless wire rope, first along the drive on the horizontal, and then up an incline increasing in sharpness till the shaft is reached, where the direction of motion becomes vertical. Near the surface, again, there is an incline, gradually leading to the level of the ground, or rather of the elevated tramway from which the stuff is to be tipped into the mill, or, if it be mullock, on to the waste heap. The return of each truck is effected along the reverse side of the endless wire-rope cable.

Ventilation is an incidental work of much importance which it becomes more practicable to carry out in a satisfactory manner when an endless system of truck conveyance has been provided, reaching from the ore-face to the mill, and thence back again. The reason is mainly that the same routes which have been prepared for this traffic are available for the supply of air and for the return current which must carry off the accumulated bad gases from the underground workings. Fans, operated by the cable at various places along the line of communication, keep up a brisk exchange of air, and the coming and going of the trucks themselves help to maintain a good, healthy atmosphere, even in the most remote parts of the mine. In very deep mines, where the heat becomes unbearable after a few minutes unless a strong wind be kept going underground, the forward and backward courses for traffic and ventilation together are specially advantageous.

Prices during the twentieth century will depend more definitely upon the cost of gold-mining than they have ever done at any former time in the world's history. In spite of all the opposition which fanaticism and ignorance could offer to the natural trend of events in the commercial and financial life of the world, the gold standard now rests on an impregnable base; and every year witnesses some new triumph for those who accept it as the foundation of the civilised monetary system. This being the case, it is obvious that the conditions affecting the production of gold must possess a very peculiar interest even for those who have never lived within hundreds of miles of any gold mine. To all intents and purposes the habit of every man is to measure daily and even hourly the value of his efforts at producing what the economist calls "utilities," against those of the gold miner.

If, therefore, the latter successfully calls to his aid mechanical giants who render his work easier and who enable him to throw into the world's markets a larger proportion of gold for a given amount of effort, the result must be that the price of gold must fall, or, in other words, the prices of general commodities must rise. If, on the other hand, all other industries have been subjected to the like improved conditions of working, the effect must be to that extent to balance the rise and keep prices comparatively steady.

From this point of view it will be seen that the interests of all those who desire to see a rise in general prices are to a large extent bound up in the improvement of methods for the extraction of gold. The question of cheap power does not by any means monopolise the data upon which such a problem can be provisionally decided; and yet it may be broadly stated that in the main the increased output of gold in the future depends upon the more economical production and application of power. Measured against other commodities which also depend mainly upon the same factor, gold will probably remain very steady; while, in contrast with those things which require for the production taste and skill rather than mere brute force or mechanical power, gold will fall in value. In other words, the classes of articles and services depending upon the exercise of man's higher faculties of skill, taste, and mental power will rise in price.

Getting gold practically means, in modern times, crushing stone. This statement is subject to fewer and fewer exceptions from one decade to another, according as the alluvial deposits in the various gold-producing countries become more or less completely worked out. A partial revival of alluvial mining has been brought about through the application of the giant dredger to cheapening the process of extracting exceedingly small quantities of gold from alluvial drift and dirt. Yet on the whole it will be found that the gold-mining industry, almost all the world over, is getting down to the bed-rock of ore-treatment by crushing and by simple methods of separation. Thus practically we may say that the cost of gold is the cost of power in those usually secluded localities where the precious metal is found in quantities sufficient to tempt the investment of capital.

From this it may be inferred that the cheap transmission of power by the electric current will effect a more profound revolution in the gold-mining industry than in almost any other. The main deterrent to the investing of money in opening up a new gold mine consists in the fact that a very large and certain expense is involved in the conveyance of heavy machinery to the locality, while the results are very largely in the nature of a lottery. When, however, the power is supplied from a central station, and when economical types of crusher are more fully introduced, this deterrent will, to a large extent, disappear. The cables which radiate from the central electric power-house in all directions can be very readily devoted to the furnishing of power to new mines as soon as it is found that the older ones have been proved unprofitable.

No one will think of carrying ore to the power when it is far more economical and profitable to carry power to the ore. In this connection the principle of the division of labour becomes very important. In its bearing upon the mining industry generally, whether in its application to the precious metals or to those which are termed the baser, and even in the work of raising coal and other non-metalliferous minerals, the fact that nearly all mines occur in groups will greatly aid in determining the separation of the work of supplying power, as a distinct industry from that of mining.

Ore-dressing is an art which was in a very rudimentary state at the middle of the nineteenth century, when the great discoveries of gold, silver and other metals began to influence the world's markets in so striking a manner. The ancients used the jigger in the form of a wicker basket filled with crushed ore and jerked by hand up and down in water for the purpose of causing the lighter parts to rise to the top, while the more valuable portions made their way to the bottom. In this way the copper mines of Spain were worked in the days of the Roman Empire, and probably the system had existed from time immemorial.

Fifty or sixty years ago the miner had got so far as to hitch his jigging basket or sieve on to some part of his machinery, generally his pumping engine, and thus to avoid the wearing muscular effort involved in moving it in the water by hand. It was not until the obvious mistake of using a machine which permitted the finest, and sometimes the richest, parts of the ore to escape had been for many years ineffectually admitted, that the "vanner," or moving endless band with a stream of water running on it, was invented with the special object of treating the finer stuff.

Jiggers and vanners form the staple of the miner's ore-dressing machinery at the present day. The efficiency of the latter class of separating machines, working on certain kinds of finely crushed ore, is already so great that it may be said without exaggeration that it could hardly be much improved upon, so far as percentage of extraction is concerned; and yet the waste of power which is involved is something outrageous. For the treatment of a thin layer of slimes, perhaps no thicker than a sixpence, it is necessary to violently agitate, with a reciprocating movement, a large and heavy framework. Sometimes the quantity of stuff put through as the result of one horse-power working for an hour is not more than about a hundredweight. The consequence is that in large mines the nests of vanners comprise scores or even hundreds of machines. When shaking tables are used, without the addition of the endless moving bands, good work can also be done; but the waste of power is still excessive.

The vanning spade and shallow washing dish are the prototypes of this kind of ore-dressing machinery. Let any one place a line of finely-crushed wet ore on a flat spade and draw the latter quickly through still water, at the same time shaking it, and the result on inspection, if the speed has not been so great as to sweep all the fine grains off the surface, will be that the heavier parts of the ore will be found to have ranged themselves on the side towards which the spade was propelled in its progress through the water. A sheet of glass serves for the purpose of this experiment even better than a metal implement; but the spade is the time-honoured appliance among miners for testing some kinds of finely crushed ore by mechanical separation.

It is to be observed that, besides the shaking motion imparted to the apparatus, the only active agency in the distribution of the particles is the sidelong movement of the spade relatively to the water. But it makes little or no difference whether the water moves sidelong on the spade or the latter progresses through the liquid; the ore will range itself accurately all the same. Consequently, if a circular tank be used, and if the water be set in rotary motion, the ore on a sheet of glass, held steady, will arrange itself in the same way. If the ore be fed in small streams of water down the inclined surfaces of sloping glass, or other smooth shelves set close to and parallel with one another near the periphery of such a vessel of moving water, the resultant motions of the heavy and of the light particles respectively, in passing down these shelves, will be found to be so different that the good stuff can be caught by a receptacle placed at one part, while the tailings fall into another receiver which is differently situated at the place where the lighter grains fall.

The main essential in this particular application of the art of vanning is simply that the water should move or drift transversely to lines of ore passing, while held in suspension with water, down a smooth sloping surface. In dealing with some very light classes of ore, and especially such as may naturally crush very fine—that is to say, with a large proportion of impalpable "slimes"—there is a decided advantage in causing the water to drift sidelong on the smooth shelf by other means than the motion in a circular tank.

Adopting nearly the form of the "side delivery manner," in which the moving band is canted to the side and the stuff runs off sideways, the sloping smooth shelf can be worked for ore separation with merely the streams of water holding the fine sand in suspension running down at fixed intervals. A glass covering is placed very close to this surface on which the streams run; and between the two is driven laterally a strong current of wind by means of a blast-fan, which causes each stream of water to drift a little sidewards, carrying with it the lighter particles, but leaving on its windward side a line of nearly pure ore. These small runlets can be multiplied, on a shelf measuring six or eight feet in length, to such an extent that the machine can put through as much ore as a dozen vanners, consuming only a mere fraction of the power necessary to drive one machine of the older type.

Cyanide solution, instead of water, is very advantageously employed for this kind of operation in the case of extracting gold from crushed ore. The method is to pump the liquid from the tanks in which it is stored and to allow it to flow back by way of the vanning apparatus, thus providing not only for catching the grains of gold by the concentrating machine, but also for the dissolving of the fine impalpable gold dust, or natural precipitate, by the action of the cyanide of potassium.

Upon the use of this latter chemical will be based the main improvements in the gold-mining industry during the twentieth century; and, conversely, the applications of the old system of amalgamating with mercury, in order to catch the golden particles, will be gradually restricted. Fine concentrators, worked with cyanide solution, perform three operations at once, namely, first, the catching of the free gold grains; second, the production of a rich concentrate of minerals having gold in association and intended for smelting; and, third, the dissolving of the finest particles by the continual action of the chemical.

In fact it is in the treatment of complex and very refractory ores generally, whether of the precious or of the baser metals, that the finer applications of the art of the ore-dresser will receive their first great impetus. The vanner, as well as the jigger, will become an instrument of precision; and in combination with rushing appliances operated by cheap power in almost unlimited quantities it will materially assist in multiplying the world's supply of metals. This again will aid in promoting the further extension of machinery. Gold will be produced in greater abundance for what is called the machinery of commerce; and the base metals, particularly the new alloys of steel and also copper and aluminium, will be more largely produced for engineering and electrical purposes.

The importation—particularly to England and Scotland—of large quantities of highly-concentrated iron ore will cause one of the first notable developments in the mining and ore-treatment of the twentieth century so far as the United Kingdom is concerned. The urgent necessity for an extension in the manufacture of Bessemer steel, and of the new and remarkable alloys in which very small quantities of other metals are employed in order to impart altogether exceptional qualities to iron, must accentuate the demand for those kinds of ore which lend themselves most readily to the special requirements of the works on hand. Hence the question of the transport of special kinds of iron ore over longer distances will have to be faced (as it has been already to a limited degree), and not only in reference to ores containing a low percentage of phosphorus and therefore exceptionally suitable for the Bessemerising process, but also in regard to ores which are amenable to magnetic separation.

Magnetite, indeed, must bulk more largely in the future as a source of iron, particularly because it is susceptible of magnetic separation, a process which as yet is only in its infancy. Containing, as it does, a larger percentage of iron than any other source from which the metal is commercially extracted, its employment as an ore results in great economy of fuel, as well as a reduction in the proportionate costs of transport. When ores of iron require to be brought from oversea places, it is obvious that those which will concentrate to the purest product possible, and which are in other respects specially applicable to the production of grades of steel of exceptional tensile strength, will have the preference.

Magnetic concentration, or the separation of an ore from the waste gangue by the attraction of powerful electro-magnets, must therefore occupy a much more prominent place in the metallurgy of the future than it has in that of the past. Not only may ironstone containing magnetite be separated from other material, but several important minerals acquire the property of becoming magnetic when subjected to the operation of roasting, sometimes through a sulphide being converted into a magnetic oxide.

By the use of powerful electro-magnets, the poles of which are brought to a point or to a nearly sharp knife-edge, the intensity of the magnetic field can be so enormously increased that even minerals which are only feebly magnetic can readily be separated by being lifted away from the non-magnetic material. In some systems the crushed ore is simply permitted to fall in a continuous stream through a strong magnetic field, and the magnetic particles are diverted out of the vertical in their descent by the operation of the magnets.

Nor is it only those minerals that actually become themselves magnetic on being roasted which can be so differentiated from the material with which they are associated as to be amenable to magnetic separation. Even differences in hygroscopic properties—that is to say, in the degree of avidity with which a mineral takes up moisture from the atmosphere—may be made available for the purpose of effecting a commercially valuable separation. This is especially the case with some complex ores in which one constituent, on being roasted, acquires a much greater hygroscopic power than the others, the grains of the crushed and roasted ore becoming damp and sticky while those of the other minerals remain comparatively dry. By mixing with an ore of this kind—after it has been allowed to "weather" for a short time—some finely-powdered magnetite the strongly hygroscopic constituents can be made practically magnetic, because the magnetic impalpable dust adheres to them, while it remains separate from the grains of the other minerals.

Hardness—as well as magnetic attraction—is a property of ore which has as yet been made available to only a very slight extent as the basis of a system of separation. If a quantity of mixed fragments of glass and plumbago be pounded together in a mortar with only a moderate degree of pressure, so as to avoid, as far as possible, the breaking of the glass, there will soon come a stage at which the softer material can be separated from the harder simply by means of a fine sieve. There are many naturally-existing mineral mixtures in the crushing of which a similar result occurs in a very marked degree; and, indeed, there are none which do not show the peculiarity more or less, because the constituents of an ore are never of exactly the same degree of hardness. When the worthless parts are the softer and therefore have the greater tendency to "slime," the ore is very readily dressed to a high percentage by means of water.

But when the reverse is the case, and the valuable constituents through their softness get reduced to a fine pulp long before the other parts, the ordinary operations of the ore-dresser become much more difficult to carry out. Most elaborate ore-reduction plants are constructed with the view to causing the crushing surfaces, whether of rolls or of jaws, to merely tap each piece of stone so as to break it in bits without creating much dust. This operation is repeated over and over again; but the stuff which is fine enough to go to the concentrator is removed by sieving after each operation of the kind; and the successive rolls or other crushers are set to a finer and finer gauge, so that there is a progressive approach to the conditions of coarse sand, which is that specially desired by the ore-dresser.

Much of this elaboration will be seen to be needless, and, moreover, better commercial results will be obtained when it is more clearly perceived that the recovery of a valuable ore in the form of a fine slime may be economically effected by the action of grinders specially constructed for the purpose of permitting the hard constituents of the ore to remain in comparatively large grains, while the other and softer minerals are reduced to fine slimes or dust. In other words, a grinding plant, purposely designed to carry out its work in exactly the opposite way to that which has been described as the system aimed at in ordinary crushing machinery, has its place in the future of metallurgy. Light mullers are employed to pound, or to press together, the crushed grains for a given length of time, and then sieving machinery completes the operation by taking out the dust from the more palpable grains.

In some cases it will be found that an improvement can be effected by bringing about the separation of a finer grade of dust than could be taken out by any kind of sieve which is commercially practicable on the large scale. This is more particularly the case in regard to sulphide ores containing very friable constituents carrying silver. A fine dry dust-separator may then be employed constructed on the principle of a vibrating sloping shelf which moves rhythmically, either in a horizontal circle or with a reciprocal motion, and which at the same time alters its degree of inclination to the horizontal. When the shelf is nearly level its vibration drives the coarser particles off; but the very finest dust does not leave it until it assumes nearly a vertical position. A large nest of similar shelves, set close to, and parallel with, one another, can separate out a great quantity of well-dried slimes in a very short space of time.

CHAPTER IX
DOMESTIC

The enormous waste involved in the common methods of heating is one of the principal defects of household economy which will be corrected during the twentieth century. Different authorities have made varying estimates of the proportion between the heat which goes up the chimney of an ordinary grate, and that which actually passes out into the room fulfilling its purpose of maintaining an equable temperature; but it cannot be denied that, at the very least, something like three-fourths of the heat generated by the domestic fires of even the most advanced and civilised nations goes absolutely to waste—or rather to worse than waste—because the extra smoke produced in creating it only serves to pollute the atmosphere. In the cities some degree of progress has been made in the introduction of heating appliances which really give warmth to a room without losing at least seventy-five per cent. of their heat; but in the country districts, where open fireplaces are the rule, it is not unusual to find that more than ninety per cent. of the heat produced behind the domestic hearth goes up the chimney.

Sentiment has had a great deal to do with retarding progress in the direction of improved house-heating appliances. For countless ages "the hearth" has been, so to speak, the domestic altar, around which some of the most sacred associations of mankind have gathered, and popular sentiment has declared that it is not for the iconoclastic inventor or architect to improve it out of existence, or even to interfere seriously with either its shape or the position in the living room from which it sheds its genial warmth and cheerfulness around the family circle. A recognition of this ineradicable popular feeling was involved in the adoption of the grate, filled with glowing balls of asbestos composition, by the makers of gas-heating apparatus. The imitation of the coal-filled grate is in some cases almost perfect; and yet it is in this close approximation to the real article that some lovers of the domestic fuel-fire find their chief objection, just as the tricks of anthropoid animals—so strongly reminiscent of human beings and yet distinct—have the effect of repelling some people far more than the ways of creatures utterly unlike man in form and feature.

Taking count of the domestic attachment to a real fuel-filled fireplace or grate as one of the principal factors in the problem of domestic heating, it is plain that one way of obviating the waste of heat which is at present incurred, without doing violence to that sentiment, is by making better use of the chimney. The hot-air pipes and coils which are already so largely used for indoor heating offer in themselves a hint in this direction. Long pipes or coils inserted in the course taken by the heated air in ascending a chimney become warm, and it is possible, by taking such a pipe from one part of the room up the passage and back again, to cause, by means of a small rotating fan or other ventilating apparatus, the whole of the air in the chamber to circulate up the chimney and back again every few minutes, gathering warmth as it goes. In this way, and by exposing as much heating surface to the warm air in the chimney as possible, the warmth derived by an ordinary room from a fuel fire can be more than doubled.

At the same time the risk of spreading "smuts" over the room can be entirely avoided first by keeping the whole length of pipe perfectly air-tight, and attaching it in such a way as to be readily removed for inspection; and, secondly, by placing the outward vent in such a position that the gentle current must mount upwards, and any dust must fall back again into a wide funnel-shaped orifice, and by covering the latter with fine wire gauze. An apparatus of this kind acts as a remover of dust from the room instead of adding any to it. One necessity, however, is the provision of motive power, very small though it be, to work the fan or otherwise promote a draught.

Electric heating is, however, the method which will probably take precedence over others in all those cases where systems are tried on their actual merits apart from sentiment or usage. The wonderful facility afforded by the electric heating wire for the distribution of a moderate degree of warmth, in exactly the proportions in which it may be needed, gives the electric method an enormous advantage over its rivals. The fundamental principle upon which heating by electricity is generally arranged depends upon the fact that a thin wire offers more electrical resistance to the passage of a current than a thick one, and therefore becomes heated. In the case of the incandescent lamp, in which the carbon filament requires to be raised to a white heat and must be free to emit its light without interference from opaque matter, it is necessary to protect the resisting and glowing material by nearly exhausting the air from the hermetically sealed globe or bulb in which it is enclosed.

But in electrical house-warming, for which a white heat is not required and in which the necessary protection from the air can be secured by embedding the conveying medium in opaque solid material, the problem becomes much simpler, because strong metallic wires can be used, and they may be enclosed in any kind of cement which does not corrode them and which distributes the heat while refusing to conduct the electric current. A network of wire, crossing and recrossing but always carrying the same current, may be embedded in plaster and a gentle heat may be imparted to the whole mass through the resistance of the wires to the electricity and their contact with the non-conducting material.

Concurrently with this method of heating there is gradually being introduced a practice of using metallic lathing for the plastering of dwelling-rooms in place of the old wooden battens generally employed for lath-and-plaster work. The solution of the practical problem which has to be faced seems to depend upon the prospect of effecting a compromise between the two systems, introducing thin resisting wire as the metallic element in such work, but making all other components from non-conducting material. In the event of any "cut-out" or "short-circuiting" occurring through accidental injury to the wall, it would be very inconvenient to be compelled to knock away the plaster. Moreover, it is not necessary for ordinary warming purposes that the whole of the wall, up to the ceiling, should be heated.

Accordingly the system which is likely to commend itself is that of constructing panels on some such principle as the one already described, and affixing them to the wall, forming a kind of solid dado from three to four feet from the floor. These can be fastened so as to facilitate removal for examination and repairs. When the current is switched on they are slowly warmed up by the heat generated through the resistance of the wires, and the air in the room is gently heated without being vitiated or deprived of its oxygen as it is by the presence of flames, whether of fuel or of gas. Warming footstools will also be provided, and a room heated in this way will be found eminently comfortable to live in.

This method of house-warming having once obtained a decided lead within the cities and other localities where a cheap electric current is available, somewhat similar systems, adapted for the heating of walls by hot air in tubes, instead of by resistant wires, will be largely adopted in the rural districts, more particularly in churches and other places of public assemblage. The progress made in this direction during the last few years of the nineteenth century is already noteworthy, but when electric-heating really gets a good chance to force the pace of improvement, the day will soon arrive when it will be regarded as nothing less than barbarous to ask people to sit during the winter months in places not evenly warmed all through by methods which result in the distribution of the heat exactly as it is wanted.

Ventilation is another household reform which will be very greatly accelerated by the presence of electric power of low cost. The great majority of civilised people, as yet, have no idea of ventilation excepting that highly unreasonable kind which depends upon leaving their houses and other buildings partly open to the outside weather. One man is sitting in church under a down draught from an open window above him, while others, in different parts of the same building, may be weltering in the heat and feeling stifled through the vitiated air. In dwelling-houses the great majority of living rooms really have no other effective form of ventilation than the draught from the fireplace. The strength of this draught, again, is regulated to a very large extent by the speed and direction of the outside wind.

In calm and sultry weather, when ventilation is most needed, the current of air from the fireplace may be very slight indeed; while in the wild and boisterous days succeeding a sudden change of weather, the living rooms are subjected to such a drop in temperature and are swept by such draughts of cold air that the inmates are very liable to catch colds and influenza. Hence has arisen in the British Islands, and in the colder countries of Europe and America, the very general desire among the poorer classes to suppress all ventilation. Rooms are closed at the commencement of winter and practically remain so until the summer season. Many people whose circumstances have improved, and who pass suddenly from ill-ventilated houses to those which have better access to the outside air, find the change so severe upon their constitutions and habits that they give a bad name to everything in the shape of ventilation. Meanwhile the dread of draughts causes people to exclude the fresh air to such an extent that consumption and many other diseases are fostered and engendered.

All this arises mainly from the very serious mistake of imagining that it is possible to move air without the exercise of force. In the case of the draught caused by a fire no doubt an active force is employed in the energy of the heated air ascending the chimney, and in the corresponding inrush. This latter is usually drawn from below the door—the very worst place from which it can be taken, seeing that in the experience of most people it is by getting the feet chilled, through draughts along the floor, that the worst colds are generally contracted. Fireplaces are not unusually regarded as a direct means for ventilation, and with regard to nearly all the devices commonly adopted in houses and public buildings, it may be said that they lack the first requisite for a scientific system of renewing the air, namely a source of power by means of which to shift it from outside to inside, and *vice versâ*. There is no direction in which a more pressing need exists for the distribution of power in small quantities than in regard to the ventilation of private and public edifices.

The circular fan, placed in the centre piece of the ceiling and controlled by an electric switch on the wall, is the principal type of apparatus applicable to the purposes of ventilation. As electric lighting of dwelling-houses becomes more common, and ultimately almost universal within cities, the practice will be to arrange for lighting and for ventilation at the same time. But, unfortunately, the current now principally employed for electric lighting and consisting of a series of impulses, first in one direction and then in the opposite, "alternating" with wonderful rapidity, is not well adapted for driving small motors of the types now in use. One improvement in domestic economy greatly needed in the twentieth century consists in the invention of a really effective simple and economical "alternate-current" motor. This is a matter which will be referred to in dealing with electrical machines. That the problem will be solved before many years have passed there is no good reason to doubt.

In the meantime many laudable endeavours are being made towards the application of the pressure from water pipes to the purpose of driving ventilating fans. The extreme wastefulness of power and of water involved in this method of dealing with the difficulty may be partially overlooked on account of the very small amounts required to produce an effect in the desired direction; and yet there is no doubt that a recognition of the wastefulness acts to some extent as a deterrent to artificial ventilation. The benefits of the system are not sufficiently obvious or showy to induce any class of people, excepting physicians and persons fully acquainted with the principles of hygiene, to sanction a material outlay upon the object. When an exactly suitable alternate-current motor has been invented the standard electric light installation will be practically one apparatus with the ventilating fan, and the cost of the latter will hardly be felt as a separate item.

In cooking there is in existing ordinary methods the same enormous waste of heat as there is in the warming of rooms. Something, no doubt, has been done in the direction of economy by the invention of new and improved forms of stoves, but a great preponderance of the heat generated in the fire of even the best stove goes up the chimney. The electric oven, as already invented, is perhaps the nearest approach to a really economical "cooker" that has yet been proposed; but even before the general adoption of such an apparatus there will be ample room for improvement in the cooking stove, first as regards insulation, and secondly in the distribution of the fuel around the objects to be heated. One principal cause of the waste that goes on arises from the fact that the fire burns away from the place at which its heat is most beneficially applied, and no means are adopted, as in the case of the candle in a carriage lamp, for keeping it up to the required level. Additions of fuel are made from the top with the immediate effect of checking the heat.

A great advance in economy of fuel will take place when the household coal intended for cooking purposes is ground up together with the proper proportions of certain waste products of chemistry, so as to make a "smouldering mixture" which can be kept regularly supplied to a shallow or thin fire box by pressure applied from beneath or at the parts farthest away from the objects to be heated. An oven, for instance, may be surrounded by a "jacket" filled with ground smouldering mixture having a non-conducting insulator outside and a connection with a chimney. The heat from the fuel is thus kept in close proximity to the objects requiring to be cooked, and comparatively small waste results.

It is by taking advantage of their superior facilities in the same direction that gas and inflammable oils have already made their mark in the sphere of domestic cookery. Regarded as fuel their initial cost may be relatively heavy; and yet, owing to their more exact method of application, they often effect a saving in the end. Not only do they bring the fire closer to the articles to be heated or cooked, but they also make it easy for the fire to be turned off or on, and this in itself is an important source of economy. Still, with the advent of cheaper and more accessible power in every centre of population, the cost of grinding coal and of mixing it in order to form a fuel comparable in respect of convenience and economy with gas and oil will be so greatly reduced that the "black diamond" will still continue to challenge its rivals in the arena of competition presented by the demands of domestic economy.

Light, as well as heat and air, requires to be evenly and equably distributed throughout the dwelling-house before anything approaching an ideal residence can be secured. As the science of hygiene advances it is demonstrated more and more clearly that sunlight—and even diffused daylight—may be used as a most effective weapon against the spread of disease. Alternations of deep gloom in the dwelling-house with the superior light resulting from brighter weather produce many kinds of nervous derangement, not the least deleterious of which arise from the unnecessary strain to which the eyesight is subjected. The promise of the future is that, through the abundance of windows provided in the walls, roofs and porches of our dwelling-houses—but all supplemented with shutters and blinds of various kinds—there shall be a possibility of regulating, far more accurately than at present, the accessibility of light from outside according to the brightness or dulness of the day.

It is hardly to be expected that many people will build "Crystal Palaces" in which to reside; but with the immense progress that is being made in the construction of dwellings with iron or steel frames, and in the adaptation of various materials so that they may serve for building purposes in conjunction with metallic frameworks, it seems clear that many roofs, as well as large portions of walls, will in future be made on the composite principle, using steel and glass. These will, to a large extent, be permanently sheltered from the direct rays of the sun when high in the heavens, by shutters constructed on the louvre principle so that they may admit the light from the sky continually, but actual rays or beams of sunlight only for a short time after sunrise and at the close of day. The ceilings, if any are provided under the roofs, will also be glazed.

The obstacles presented in the way of such a reform in a city like London may at first sight seem so serious as to be practically insuperable. Long rows of three or four storied houses certainly offer but few facilities for the admission of light through the roofs of any but the rooms on the top floors, and yet it is in the dwelling-houses of this type that the depression caused by gloom and the absence of light during the hours of day are most severely felt as a source of nervous depression. Evolution in a matter of this sort will take place gradually and along the line of least resistance. Portions of courts, areas and yards will be glazed over in the way described; and it will be found that those rooms which are thus enclosed and sheltered from the wind and rain, but left open to the daylight, constitute the most cheerful sitting places in the houses. Then, as rebuilding and alterations proceed, many houses will gradually be remodelled—at least as regards some of their rooms—in the same direction. Physicians will become increasingly insistent on the necessity for admitting plenty of light into the abodes of the sick, more particularly of families inclined towards consumption.

A very large trade will spring up during the twentieth century in household cooling apparatus for use in hot climates. The colonial expansion towards which all European races are now tending inevitably means that very many thousands of persons whose ancestors have been accustomed to life in cold or temperate climates, will be induced to dwell in the dry and warm, or in the humid tropical regions of the earth. It will be an important task of the British, Continental and American machinists of the twentieth century to turn out convenient pieces of apparatus which shall be available for ventilating houses, especially during the night, and for reducing the temperature in them to something approaching that which is natural to the inmates. The old clumsy punkah will be replaced by circular fans keeping up a gentle current of air with a minimum of noise or annoyance of any kind.

At present it is only in specially favoured circumstances that these quiet-working circular punkahs can be actuated by mechanical force, that is to say where a prime motor, or an electric current, or a reticulated water supply for driving a suitable machine may be at hand. In other situations the use of compressed air or gas may be resorted to, and for this purpose small capsules, similar to those already introduced for making soda water by the liberation of compressed carbonic acid gas, will be found handy. For a very small sum of money the householder will be able to purchase a sufficient number of capsules to ensure motive power for his fan during a week of hot nights.

A convenient form of small motor suitable for being driven by compressed air or gases in this way is one in which a diminutive turbine or other wheel is set at the bottom of a thin tube of mercury. The capsule, being fastened to the lower end of this apparatus, liberates at very short intervals of time bubbles of air or gas, which, in the upward ascent, drive the wheel. The arrangement depends upon the fact that a stream of gas ascending in a heavy liquid behaves in the same way as a stream of water descending by its own weight and turning a water-wheel. It supplies what is perhaps the simplest and most inexpensive small motor available for the lightest domestic work to which a gentle but continuous source of power is applicable.

For actually cooling the air, as well as keeping it in motion, similar devices will be resorted to, with the addition of the circulation of the current of air through coils of pipes laid under the surface of the ground. In this way householders will have all the advantages of living in cool underground rooms without incurring the discomforts and dangers which are often inseparable from that mode of life. In the coastal regions, which usually have the most trying climates for Europeans living in tropical countries, a method of cooling the houses will be based on the fact that at moderate depths in the sea the prevailing temperature is a steady one, not much above the freezing point of water. Almost every seaport town within the tropics—where white residents in their houses swelter nightly in the greatest discomfort from the heat—is in close proximity to deep ocean water, in which, at all seasons of the year, the regular temperature is only about thirty-four degrees Fahr. The cost of steel piping strong enough to withstand the pressure of the water in places which possess absolutely the coolest temperature of the ocean would be very heavy; but, on the other hand, the actual reduction of heat demanded for the satisfactory cooling of the air in a dwelling-room is not by any means great, and at quite shallow depths the heat of the air can be satisfactorily abstracted by the sea water surrounding coils of pipes.

Even in colder climates it seems likely that similar systems will be found useful in producing a preliminary reduction in the temperature of the air employed in keeping fresh foodstuffs such as meat, fruits and vegetables. Fruits especially, when placed in suitable receptacles, and stored at temperatures quite steady at about the freezing point of water, will not only be readily kept on land from one season to another, but will be transported to markets thousands of miles distant from the growers, and sold in practically the same condition as if they had just been picked from the trees. During the twentieth century the proportion of the fruit eaters among the peoples of the great manufacturing countries will be very largely augmented, and this result will be brought about mainly through the instrumentality of methods of keeping perishable produce free from deterioration by maintaining it almost at the freezing point—a temperature at which, under suitable conditions as regards exclusion of moisture, and steadiness of hygrometric pressure, the germs of decay in food are practically prevented from coming to maturity.

For the cooling of dwelling-rooms in places distant from the sea, various systems, depending upon the supply of dry cold air from central stations through pipes to the dwellings of subscribers, will no doubt be brought into operation. This, however, will only be practicable in the more populous localities having plenty of residents ready to contribute to the expense. For more isolated houses the cooling and ventilating apparatus of the future may be a modification of the "shower-blast" which has been successfully adapted to metallurgical purposes. When downward jets of water, as in a shower-bath, are enclosed in a large pipe connected horizontally with a room but having facilities for the escape of the water underneath, a strong draught of cool air is created, and the prevailing temperature is quickly reduced. An apparatus of this kind may be intended for application either to the ventilators or to the windows of rooms.

Lifts for conveying persons from one storey of a building to another will probably undergo a considerable amount of modification during the next few years. The establishment of central electric stations and the distribution of electricity for lighting and for power will offer a very great premium upon the preference for electric motors for lifts. As soon as a maximum of efficiency, combined with the minimum of cost, has been attained, there will be a demand for the introduction of lifts in positions where the traffic is not large enough to warrant the constant presence of an attendant. In fact the desire will be for some kind of elevator which shall be just as free to the use of each individual as is the staircase of an ordinary house.

For this purpose, inclined planes having moving canvas or similar ramps will be extensively brought into use. The passenger steps upon what is practically an endless belt having suitable slats upon it to prevent his foot from slipping, and, as the hand-railing at the side of this moves concurrently, he is taken up, without any effort, to the landing on which he may alight quite steadily. When this idea, which has already been brought into operation, has been more fully developed, it will be seen that a large circular slowly-revolving disc, set at an angle and properly furnished, will supply a more convenient form of free elevator. One side will be used by those who are going up and the other by those who wish to come down. The "well" of the staircase for such a lift is made in elliptical form, like the shadow projection of a circle. Steps can be provided so that, when not in motion, the lift will be a staircase not differing much from the old style.

CHAPTER X
ELECTRIC MESSAGES, ETC

The telegraphic wire in the home and street will fulfil a very important part in the economy of the twentieth century. For conveying intelligence, as well as for heating, cooking and lighting, the electric current will become one of the most familiar of all the forces called in to assist in domestic arrangements. The rapidity with which the electric bell-push has taken the place of the old-fashioned knocker and the bell-hanger's system affords one indication of the readiness with which those forms of electric apparatus which are adapted to all the purposes of communicating and reminding will recommend themselves to the public during the twentieth century.

In another direction the eagerness with which every advance in the telephone is hailed by the people may well offer an augury of rapid progress in the immediate future. In this department invention will aim just as much at simplification as at elaboration; and some of the pieces of domestic electrical apparatus universally used during the twentieth century will be astonishingly cheap.

The call to awake in the morning will, in cities and towns, be made by wireless telegraphy, which will also be used for the purpose of regulating the domestic clocks, so that if desired any suitable form of clock alarm may be used with the most perfect confidence. A tentative system of this kind has been adopted in connection with certain telephone exchanges, in which special officers are told off whose duty it is to call those subscribers who have paid the small fee covering the expense. These officers are required to time their intimations according to the previously expressed wishes of subscribers. This kind of service, as well as the regulation of the household clock, is eminently a department of domestic economy in which wireless telegraphy will prove itself useful, because it does not demand that a subscriber shall have gone to the expense of installing a wire to his house and of paying a rent or fee for the use of one.

The clock controlled by wireless telegraphy will doubtless undergo a rapid development from the time when it is first introduced. Practically the same principles which enable the electrician to utilise the "Hertzian waves," or ether vibrations, for the purpose of setting a clock right once a day, or once an hour, will permit of an impulse, true to time, being sent from the central station every second, or every minute, and when this has been accomplished it will be seen that there is no more use for the maintenance of elaborate clockworks at any place excepting the central station. The domestic clock will, in fact, become mainly a "receiver" for the wireless telegraphic apparatus, and its internal mechanism will be reduced, perhaps, to a couple of wheels, which are necessary to transmit the motion of a minute-hand to that which indicates the hours.

The fire-alarm of the future must be very simple and inexpensive in order to ensure its introduction, not only into offices and warehouses but also into shops and houses. The fire-insurance companies will very shortly awake to the fact that prompt telegraphic alarm in case of fire is worth far more than the majority of the prohibitions upon which they are accustomed to insist by way of rendering fires less likely. The main principles upon which the electric fire-alarm will be operated have already been worked out and partially adopted. In the system of fuses and cut-outs used in connection with electric lighting, the methods of preventing fire due to the development of excessive heat have been well studied. But simplification is particularly required in the case of those fire-alarms which are to be useful for giving intimation of a conflagration from any cause arising.

As the telegraphic and telephonic wires are extended so as to traverse practically all the streets of every city, the fire-insurance companies will find it to their advantage to promote a simple plan, depending on the use of a combustible thread passing round little pulleys in the corners of all the rooms and finally out to the front, where an electrical "contact-maker" is fixed, so that on the thread being burnt and broken at any point in its circuit, an electric message will be at once sent along the nearest wire to the fire-brigade station and a bell set ringing both inside and outside the premises.

Somewhat similar systems will be used for checking the enterprises of the burglar. The best protected safes of the future will be enmeshed in networks of wires encased in some material which will render it impossible to determine their positions from the outside. These wires will be so related to an electric circuit that the breaking of any one of them, at any part of its course, will have the effect of ringing a bell and giving warning at the police station, as well as at other places where potential thief-catchers may be on hand. For doors and windows very simple contact devices have already been brought out, but the principal objection to their general adoption arises from the fact that so very many houses remain unconnected with any telephone system which may be made available for calling the police. Even were all houses connected it is true that in some instances attempts might be made to cut the wires when a raid was in contemplation, but the risk of discovery in any such operation would prove a very powerful deterrent. In fact the telephone wire, more than any other mechanical device, is destined to aid in "improving" the burglar out of existence.

With the indefinite multiplication of telephone subscribers at very cheap rates, there will come a powerful inducement towards the invention of new appliances for rendering the subscriber independent of the attention of officers at any central exchange. The duty of connecting an individual subscriber with any other with whom he may desire to converse is, after all, a purely mechanical one, and eminently of a kind which, by a combination of engineering and electrical skill, may be quite successfully accomplished. In the apparatus which will probably be in use during the twentieth century, each subscriber will have a dial carrying on its face the names and numbers of all those with whom he is in the habit of holding communication. This will be his "smaller dial," and beside it will be another, intended for only occasional use, through which, by exercising a little more patience, he may connect himself with any other subscriber whatever. Corresponding dials will be fixed in the central office.

Under this system, when the subscriber desires to secure a connection, he moves a handle round his dial until the pointer in its circuit comes to the desired number. An electrical impulse is thus sent along the wire to the central station for every number over which the pointer passes, and the corresponding pointer or contact-maker at the central station is moved exactly in sympathy. When the correct number is reached the subscriber is in connection with the person with whom he desires to converse. If, however, the latter should be already engaged, a return impulse causes the bell of the first subscriber to ring. Of course the prime cost of installing such a system as this will be greater than in the case of the simple hand-connected telephones; but the two systems can be used conjointly, and the immense convenience, especially to large firms, of being able to go straight to the parties with whom they wish to communicate, will induce many of them to adopt the automatic apparatus as soon as it has been perfected.

Wireless telephony must come to the front in the near future, but at first for only very special purposes. The prospect of the profits that would be attendant on working up a business unhampered by the heavy capital charges which weigh upon the owners of telephone wires must stimulate inventive enterprise to a remarkable degree in this particular line. The main difficulty, however, in the application of the system to general purposes will lie in the need for an ingenious but simple means for enabling one subscriber to call another.

For this purpose probably the synchronised clock system already referred to will be found essential, each office or house being furnished with a timekeeper of this type kept in constant agreement with a central clock, and so arranged that only when the ethereal electrical impulse is given at a certain fixed point in the minute, will any particular subscriber's bell be rung. This may be effected by some such arrangement as a revolving drum, perforated at a different part of its periphery for each individual subscriber, and capable of permitting the electrical contact which makes a magnet and rings the bell only at the fraction of a moment when the subscriber's slot passes the pointer.

This will mean, of course, that only at a certain almost infinitesimally small space of time in the duration of each minute will it be possible to call any particular subscriber, or rather to release the mechanism which will set his bell ringing for perhaps a minute at a time. In the presence of unscrupulous competition, resulting in the flinging out of Hertzian wave vibrations promiscuously, for the purpose of destroying a rival's chances of obtaining satisfactory connections, it would be necessary to make rather more complicated arrangements of a nature analogous to those of the puzzle lock. Instead of one impulse during the minute, two or three would be required, in order to release the mechanism for ringing any subscriber's bell; and no ring would take place unless the time-spaces between these impulses were exactly in accordance with the agreed form, which might be varied at convenient intervals.

Yet in the cases in which wireless telephony and telegraphy are taken up by local public authorities having power to forbid any one playing "dog in the manger," by preventing useful work by others while failing to promote it himself, the simpler system of wireless telephone call will be practicable. With the advance of municipalisation, and of intelligent collectivism generally, enterprises of public utility will be guarded from mere cut-throat commercial hostility much more sedulously in the twentieth century than they have been in the past.

A great multitude of new applications of the telegraphic and telephonic systems will be introduced in the immediate future. Not only will those subscribers who are connected by wire with central stations have the advantage of being called at any hour in the morning according to their intimated wishes, but such services as lighting the fires in winter mornings, so that rooms may be fairly warmed before they are entered, will be performed by electric messages sent from a central station.

Drawings will also be despatched by telegraph. For such purposes as the transmission of sketches from the scene of any stirring event, the first really practical application of drawing by telegraph will probably depend upon the use of a large number of code words divided into two groups, each of which, on the principles of co-ordinate geometry, will indicate a different degree of distance from the base line and from the side line respectively, so that from any sketch a correct message in code may be made up and the drawing may be reconstructed at the receiving end. Illustrated newspapers will in this way obtain drawings exactly at the same time as their other messages, and distant occurrences will be brought before the public eye much more vividly and more correctly than has ever hitherto been practicable.

For special objects, also, photographs can be sent by telegraph through the use of the photo-relief in plaster of Paris, or other suitable material, which travels backwards and forwards underneath a pointer, the rising and falling of which is accurately represented by thick and thin lines—or by the darker and lighter photographic printing of a beam of light of varying intensity—at the other end, so that a shaded reproduction of the photograph is produced. Relief at the sending end is in this way translated into darkness of shade at the receiving end. Any general expansion of this system, if it comes, will necessarily be postponed till long after the full possibilities of the codeword plan have been exploited, because the latter works in exactly with the ordinary methods for sending telegraphic matter.

The keen competition between submarine and wireless telegraphy will be one of the most exciting contests furnished by electrical progress in the first quarter of the new century. Attention will be devoted to those directions on the surface of the globe in which it is possible to send messages almost entirely by land lines, and to bridge over comparatively small intervals of space from land to land by wireless telegraphy. Thus the Asiatic and Canadian route may be expected shortly to enter into competition with the Atlantic cables in telegraphic business to the United States; while Australia will be reached *viâ* Singapore and Java.

A great impetus will be given to the wireless system as a commercial undertaking when arrangements have been perfected for causing the receiver at any particular station to translate its message into a form suitable for sending automatically. When this has been done, many of the wayside stations will be almost entirely self-working, and messages, indeed, may be despatched from island to island, or from one floating station to another across the Atlantic itself.

Another requirement for really cheap telegraphy on the new system is a more rapid method of making the letters or signals. The irregular intervals at which the sparks from the coil of the transmitter fly from one terminal to the other render it impossible to split up the succession of flashes into intervals on the dot-and-dash principle, without providing for each dot a much longer period of time than is required for the transmission of messages on land lines. In fact the need for going slowly in the sending of the message is the principal stumbling-block which disconcerts ordinary telegraphic operators when they come to try wireless telegraphy. For remedying this defect the most hopeful outlook is in the direction of a multiplication of the pieces of apparatus for spark-making and the combining of pairs of them in such a way that, whenever the first one fails during an appreciable interval of time to emit a spark, the second is called into requisition. In this way a constant stream of sparks may be ensured, without incurring the risk of running faster than the coil will supply the electrical impulses necessary for the transmission of the message.

Increased rapidity in land telegraphy by the ordinary system of transmission by wire, and facility in making the records at the receiving end in easily read typewriting—these are two desiderata which at the close of the nineteenth century have been almost attained, but which will take some time to introduce to general notice. In the commercial system of the twentieth century the merchant's clerk will write his messages on a typewriter which perforates a strip of paper with holes corresponding to the various letters, while it sets down in printing, on another strip, the letters themselves. The latter will be kept as a record, but the former will be taken to the telegraph office and put through the sending machine without being read by the operator. The message will print itself at the other end and wrap itself up in secret, nothing but the address being made visible to the operator.

For the use of the general public who are not possessed of the special apparatus necessary to perforate the paper another system is available. Sets of movable type may be provided at the telegraph office in small compartments, the letters being on one side and indentations corresponding to the required perforations being cut or stamped into the other sides of the movable pieces. The sender of a message will set it up in a long shallow tray or "galley" like those used by printers, and he will then turn the faces of the letters downwards and see the whole passed through the machine without being read by the operator; after which he can distribute the letters if he chooses. In this way telegraphy will gradually become at once far more secret and far cheaper than it is at present, and a large amount of correspondence which at present passes through the post will be sent along the wire.

Many merchants will have their telephonic apparatus fitted with arrangements for setting up type or perforating strips of paper, as already described; and also with receiving apparatus for making the records in typewriting. If they fail to find a subscriber or correspondent on hand at the time when he is wanted, they can write a note to him which he will find hanging on a paper strip from his telephone when he returns. Another mode of accomplishing a somewhat similar result is to provide the telephone receiver itself with a moving strip of steel, which, in its varying degrees of magnetisation, records the spoken words so that they will, at some distance of time, actuate the diaphragm of the receiver and emit spoken words. The degree of permanency which can be attained by this system is, of course, a vital point as regards its practical merits.

Still unsolved electrical problems are the making of a satisfactory alternate current motor suitable for running with the kind of currents generally used for electric lighting purposes—the utilisation of the glow lamp having a partial vacuum or attenuated gas for giving a cheap and soft light somewhat on the principle of the Geissler tube—and last, but not least, the direct conversion of heat into electricity.

With regard to the first-mentioned, the prospects have been materially altered by a discovery announced at the New York meeting of the American Association for the Advancement of Science within a few weeks of the close of the nineteenth century. The handy and effective alternate current motor indeed seemed then as far distant as it had been in 1896, when Sir David Salomons remarked, in his work on *Electric Light Installations* (vol. ii., p. 97): "No satisfactory alternate current motor available on all circuits exists as yet, although," he added later, "the demand for such an appliance increases daily". It seems, however, that electricians have been looking in the wrong direction for the solution of using the same wire for alternate current lighting and for motive power at the same time. Professor Bedell, of Cornell University, announced at the New York meeting referred to his discovery of the important fact that when direct and alternate currents are sent over the same line each behaves as if the other were not there, and thus the same line can be used for two distinct systems of transmitting electrical energy. No time will be lost in putting this announcement to the test, not only of scientific but also of practical verification, and the probability is that all electric lighting stations in the twentieth century will contain not only dynamos of one type for the supply of light, but also direct current generators for transmitting power in all directions over the same cables.

The glow lamp having no carbon filament, but setting up a bright light with only a fraction of the resistance presented by carbon, would, if perfected, render electric lighting by far the cheapest as well as the best method of illumination. Tentative work has indicated a high degree of probability that success will be achieved, and the glowing bulb is at any rate a possibility of the future which it will be well to reckon with.

In reference to the conversion of heat into electricity without the intervention of machinery to provide motion, and thus to cause magnetic fields to cross one another, very little promise has yet been shown of any fundamental principle upon which a practical apparatus of the kind could be based. The electrician who works at this problem has to begin almost *de novo*, and his task is an immensely difficult one, although on every ground of analogy success certainly looks possible. In the meantime, as has already been indicated, the steam turbine and dynamo combined, working practically as a single machine for the generation of electricity, offers practically the nearest approach to direct conversion which is yet well in sight.

CHAPTER XI
WARFARE

The last notable war of the nineteenth century has falsified the anticipations of nearly all the makers of small arms. The magazine rifle was held to be so perfect in its trajectory, and in the rapidity with which it could discharge its convenient store of cartridges in succession, that the bayonet charge had been put outside of the region of possibility in warfare. Those who reasoned thus were forgetting, to a large extent, that while small arms have been improving so also has artillery, and that a bayonet charge covered by a demoralising fire of field-pieces, mortars, and quick-firing artillery is a very different thing from one in which the assailants alone are the targets exposed to fire. Given that two opposing armies are possessed of weapons of about equal capacity for striking from a distance, they may do one another a great deal of harm without coming to close quarters at all. Yet victory will rest with the men who have sufficient bravery, skill and ingenuity to cross the fire-zone and tackle their enemies hand to hand.

Smoke-producing shells and other forms of projected cover, designed to mask the advance of cavalry and infantry, will greatly assist in the work of rendering this task of crossing the fire-zone less dangerous, notwithstanding any possible improvement that may be effected in the magazine-rifle. Already it has been observed that much of the surprise and confusion which terrifies those who have no bayonets, when subjected to a cannonade and at the same time brought face to face with a bayonet charge, arises from the fact that they cannot see to shoot straight, owing to the haze produced by the smoke and its blinding effects upon the eyes.

Special smoke-producing shells, made for the express purpose of covering a charge, will soon be evolved from the laboratory of the chemist in pursuance of this clue. In addition to shells and other missiles, small pieces of steel-piping will be projected by mortars into the fire-swept zone, in order to supplement the defects of natural cover which, of course, are nearly always as great as possible, seeing that the ground has generally been selected by the side against which the attack is being directed.

The task of enabling a rifleman to shoot straight has been taken up with extraordinary zeal and ability compared with the amount of skill and effort devoted to the corresponding or opposing object of spoiling his aim and preventing him from getting a shot in. When this latter has been to some extent accomplished, mainly by the agency of artillery, the bayonet and other weapons for use at close quarters will once more be in the ascendant. Thin shields of hard steel will be affixed to the rifles of the attacking party, so as to deflect the bullets wherever possible.

This baffling of the rifleman by the artillery supporting the cavalry and bayonet charge will produce momentous changes, not only in the future of war, but also in that of international relations. Anything which tends to discount the value of personal bravery and to elevate the tactics of the ambuscade and the sharp-shooting expedition gives, *pro tanto*, an advantage to the meaner-spirited races of mankind, and places them more or less in a position of mastery over those who hold higher racial traditions. The man who will face the risk of being shot in the open generally belongs to a higher type of humanity than he who only shoots from behind cover.

Moreover, the nations which have the skill and ingenuity to manufacture new weapons of self-defence belong to a higher class than those which only acquire advanced warlike munitions by purchase. One of the early international movements of the twentieth century will be directed towards the prohibition of the sale of such weapons as magazine-rifles, quick-firing field guns, and torpedoes to any savage or barbarous race. It will be accounted as treason to civilisation for any member of the international family to permit its manufacturers to sell the latest patterns of weapons to races whose ascendency might possibly become a menace to civilisation. As factors in determining the survival of the fittest, the elements of high character, bravery, and intellectual development must be conserved in their maximum efficiency at all hazards.

Another potent element in the safeguards of civilisation may be seen in the increased effectiveness of weapons for coastal defence. The hideous nightmare of a barbarian irruption, such as those which almost erased culture and intellect from the face of Europe during the dark ages of the fourth, fifth and sixth centuries, may occasionally be seen exercising its influence in the pessimistic writings which are from time to time issued from the Press predicting the coming ascendency of the yellow man.

However the case may be in regard to nations which are accessible by land to the encroachments of the Asiatic, there is no doubt that those countries which are divided off by the sea have been rendered much more secure through the rapid advances which have been made in the modern appliances for defending coasts and harbours. In naval tactics, also, it will be more and more clearly seen that to possess and defend the harbours where coaling can be carried out is practically to possess and defend the trade of the high seas; and the essence of good maritime policy will be to so locate the defended harbours that they may afford the greatest amount of protection, having in view the harm that may be done by an enemy's harbours in the vicinity.

The most effective naval weapon in the future will undoubtedly be the torpedo, but, like the bayonet, it requires to be in the hands of brave men before its value as the ultimate arbiter of naval conflict can be demonstrated. Much fallacious teaching has arisen from what has been called the lessons of certain naval wars which occurred on the coasts of South America and China—international embroilments in which mercenaries, or only half-trained seamen and engineers, were engaged. On similar fallacious grounds it was argued that the magazine-rifle had put the bayonet out of the court of military arbitrament, and the South African war has proved conclusively how erroneous was that idea. The use of the torpedo-boat and of the weapons which it carries must always demand, like that of the bayonet, men of the strongest nerve, and of the greatest devotion to their duty and to their country.

Fifty miles an hour is a rate which is already in sight as the speed of the future torpedo-boat, the first turbine steamer of the British Navy having achieved forty-three miles an hour before the end of the nineteenth century. It should be distinctly understood, however, that such a speed cannot be kept up for any great length of time and that long voyages are out of the question. The rôle of the turbine torpedo-boat will be to "get home" with its weapon in the shortest practicable time. Hence its great value for the defence of harbours by striking at distances of perhaps two or three hours' steaming.

On the high seas the battle-ships, which will virtually be the cruisers of the future, will be provided with turbine torpedo-boats, carried slung in convenient positions and ready at short notice to be let slip like greyhounds. During the hazardous run of the torpedo-boat towards the enemy, various devices will be employed for the purpose of baffling his aim, such for instance as the emission of volumes of smoke from the bows and the erection of broad network blinds covering the sight of the little craft, but capable of being shifted from side to side, so that the enemy's marksmen may never know exactly what part of the object in sight is to be aimed at. The torpedo will be carried on a mast, which at the right moment can be lowered to form a projecting spar like a bowsprit; and the explosion that will take place on its impact with the enemy's hull will be enough to blow a fatal breach in any warship afloat.

For harbour defence and the safety of the battle-ship the wire-guided and propelled torpedo will form a second line behind the fast torpedo-boat. This type of weapon strikes with more unerring accuracy than any other yet included in the armoury of naval warfare, because it is under the control of the marksman from the time of its launching until it fulfils its deadly mission. Its range, of course, is strictly limited; but it may be worked to advantage within the distances at which the best naval artillery can be depended upon to make good practice.

The least costly and the lightest form is that in which the backward pulling of two wires, unwinding two drums on the torpedo, actuates two screws at greater or less speeds according to the rapidity of the motion imparted, any advantage of speed in one screw over the other being responded to by an alteration in the direction taken by the weapon. The torpedo may be set so as to dive from the surface at any desired interval; but, of course, an appearance in the form of at least a flash is necessary to enable the operator to judge in what direction he is sending his missile. Small torpedo-boats, not manned but sticking to the surface, may be used in the same manner. Each one no doubt runs a very great risk of being hit by shot or shell aimed at them; but out of half a dozen, discharged at short intervals, it would be practically impossible for an enemy to make certain that one at least did not find its billet.

The submarine boat will have some useful applications in peace; but its range of utility in warfare is likely to be very limited. It is hopeless to expect the eyes of sailors to see any great distance under the water; therefore the descent must be made within sight of the enemy, who has only to surround himself with placed contact-torpedoes hanging to a depth, and to pollute the water in order to render the assault an absolutely desperate enterprise.

Military aeronautics, like submarine operations in naval warfare, have been somewhat overrated. Visions of air-ships hovering over a doomed city and devastating it with missiles dropped from above are mere fairy tales. Indeed the whole subject of aeronautics as an element in future human progress has excited far more attention than its intrinsic merits deserve.

A balloon is at the mercy of the wind and must remain so, while a true flying machine, which supports itself in the air by the operation of fans or similar devices, may be interesting as a toy, but cannot have much economical importance for the future. When man has the solid earth upon which to conduct his traffic, without the necessity of overcoming the force of gravity by costly power, he would be foolish in the extreme to attempt to abandon the advantage which this gives him, and to commit himself to such an element as the air, in which the power required to lift himself and his goods would be immeasurably greater than that needed to transport them from place to place.

The amount of misdirected ingenuity that has been expended on these two problems of submarine and aerial navigation during the nineteenth century will offer one of the most curious and interesting studies to the future historian of technological progress. Unfortunately that faculty of the constructive imagination upon which inventive talent depends may too frequently be indulged by its possessor without any serious reference to the question of utility. Fancy paints a picture in which the inventor appears disporting himself at unheard-of depths below the surface of the sea or at extraordinary heights above the level of the land, while his friends, his rivals, and all manner of men and women besides, gaze with amazement! Patent agents are only too well aware how often an inordinate desire for self-glorification goes along with real inventive talent, and how many of the brotherhood of inventors make light of the losses which may be inflicted upon trusting investors so long as they themselves may get well talked about.

Nations may at times be infected with this unpractical vainglory of inventiveness; and on these occasions there is need of all the restraining influence of the hard-headed business man to prevent the waste of enormous sums of money. The idea that military ascendency in the future is to be secured by the ability to fly through the air and to dive for long distances under the water has taken possession of certain sections in France, Germany, Russia, Great Britain and the United States. Large numbers of voluble "Boulevardiers" in Paris have, during the last years of the nineteenth century, made it an article of their patriotic faith that the future success of the French navy depends upon the submarine boat. The question as to what an enemy would do with such a boat in actual warfare seems hardly ever to occur to them; and, indeed, any one who should venture to put such a query would run the risk of being set down as a traitor to his country!

More important to the student of the practical details of naval preparation is the great question as to the point at which the contest between shot and armour will be brought to a standstill. That it cannot proceed indefinitely may be confidently taken for granted. The plate-makers thicken their armour while the gun-makers enlarge the size and increase the penetrative power of their weapons, until the weight that has to be carried on a battle-ship renders the attainment of speed practically impossible.

Meanwhile there is going forward, in the hull of the vessel itself, a gradual course of evolution which will eventually place the policy of increasing strength of armour and of guns at a discount. The division of the air-space of a warship into water-tight compartments will doubtless prove to be, in actual naval conflict, a more effectual means of keeping the vessel afloat than the indefinite increase in the thickness and consequent weight of her armour.

The most advanced naval architects of modern times are bestowing more and more attention upon this feature, as affording a prospect of rendering ships unsinkable, whether through accidents or through injury in warfare. No doubt, for merchant steamers, it will be seen that development along the lines already laid down in this department will suffice for all practical purposes. The water-tight bulkheads, with readily closed or automatically shutting doorways, ensure the maintenance of buoyancy in case of any ordinary accident from collision or grounding, while the duplication of engines, shafts and propellers—without which no steamship of the middle twentieth century will be passed by marine surveyors as fit for carrying passengers on long ocean voyages—will make provision against all excepting the most extremely improbable mishaps to the machinery.

If the numerical estimate of the chance of the disablement of a single engine and its propeller during a certain voyage be stated at one to a thousand, then the risk of helplessness through the break down of both systems in a vessel having twin screws and entirely separate engines will be represented by the proportion of one to a million. This mode of reckoning, of course, assumes that the two systems could be made absolutely independent in relation to all possible disasters; and some deduction must be made on account of the impossibility of attaining this ideal. Yet it is evident that when every practicable device has been adopted for rendering a double accident improbable the chances against such a disaster will not be far from the proportion stated.

When we come to consider the evolution of the warship as compared with that of the merchant steamer, we are at once confronted with the fact that the infliction of injury upon the boilers, the engine, or the propellers of a hostile vessel is the great object aimed at by the gunners. The evolution of the warship in the direction of ensuring safety, therefore, will not stop at the duplication of the engines, boilers and propellers. In fact it must sooner or later be apparent that the interests of a great naval power demand the working out of a type of warlike craft that shall be almost entirely destitute of armour, but constructed on such a principle—both as to hull and machinery—that she can be raked fore and aft, and shot through in all directions without becoming either water-logged or deprived of her motive power.

A torpedo-boat built on this system may consist essentially of a series of steel tubes of large section grouped longitudinally, and divided into compartments like those of a bamboo cane. Each of these has its own small but powerful boilers and engines, and each its separate propeller at the stern. Care also is taken to place the machinery of each tube in such a position that no two are abreast. In fact, the principle of construction is such as to render just as remote as may be the possibility of any shot passing through the vessel and disabling two at the same time.

If a boat of this description has each tube furnished not only with a separate screw at the stern, but also with a torpedo at the bows, it can offer a most serious menace to even the most powerful battle-ship afloat, because its power of "getting home" with a missile depends not upon its protective precautions, but upon an appeal to the law of averages, which makes it practically impossible for any gunners, however skilful, to disable all its independent sections during the run from long range to torpedo-striking distance. The attacked warship is like an animal exposed to the onslaught of one of those fabled reptiles possessing a separate life and a separate sting in each of its myriad sections; so that what would be a mortal injury to a creature having its vital organs concentrated in one spot produces only the most limited effect in diminishing its strength and powers of offence.

Or this class of naval fighter may be regarded as a combined fleet of small torpedo-boats, bound together for mutual purposes of offence and defence. Singly, they would present defects of coal-carrying capacity, sea-going qualities, and accommodation for crew which would render them comparatively helpless and innocuous; but in combination they possess all the travelling capacities of a large warship, conjoined with the deadly powers at close quarters of a number of torpedo boats, all acting closely in concert upon a single plan.

The chief naval lesson taught during the Spanish-American War was the need for improving the sea-going qualities of the torpedo-boat before it can be regarded as a truly effective weapon in naval warfare. It was announced at one stage that if the Spanish torpedo-boat fleet could have been coaled and re-coaled at the Azores, and two or three other points on the passage across to America, it might have been brought within striking distance of the United States cruisers operating against Santiago. This hypothetical statement provided but cold comfort for the Spaniards, who had been persuaded to put so much of their available naval strength into a type of craft utterly unsuited for operations complying with the first great requirement of naval warfare, namely, that the proper limit of the campaign coincides with the shores of the enemy's country.

But when the naval architect and the engineer have evolved a class of torpedo-using vessel which can both travel far and strike hard, and which, moreover, can stand a few well-directed shots penetrating her without succumbing to their effect, a new era will have been opened up in naval warfare—an era of high explosive weapons requiring to strike home with dash and bravery in spite of risk from shot and shell; but, like the bayonet on land, capable of overthrowing all war-machines which can only strike from a considerable distance.

CHAPTER XII
MUSIC

A perfect *sostenuto* piano has been the dream of many a musician whose ardent desire it was to perform his music exactly as it was written. A sustained piano note is, indeed, the great mechanical desideratum for the music of the future. In music, as at present written and published for the piano, which is, and must continue to be, the real "King of Instruments," there is a good deal of make-believe. A long note—or two notes tied in a certain method—is intended to be played as a continued sound, like the note of an organ; whereas there is no piano in existence which will produce anything even approximately approaching to that effect. The characteristic of the piano as an instrument is *percussion*, producing, at the moment of striking the note, a loud sound which almost immediately dies away and leaves but a faint vibration.

The phonographic record of a pianoforte solo shows this very clearly to the eye, because the impression made by a long note is a deeply-marked indentation succeeded by the merest shallow scratch—not unlike the impression made by a tadpole on mud—with a big head and an attenuated body. Every note marked long in pianoforte music is therefore essentially a *sforzando* followed by a rapid *diminuendo*. Anything in such music marked as a long note to be sustained *crescendo*—the most thrilling effect of orchestral, choral, and organ music—is necessarily a sham and a delusion.

The genius and skill which have enabled the masters of pianoforte composition not only to cover up this defect in their instrument, but even to make amends for it, by working out effects only suitable for a percussion note, present one of the most remarkable features of musical progress in the nineteenth century. So notable is that fact in its relation to the pianoforte accompaniments of vocal music, that it seems open to question whether, even in the presence of a thoroughly satisfactory *sostenuto* piano, much use would for many years be made of it for this particular purpose. The effects of repeated notes succeeding one another with increasing or decreasing force, and of *arpeggio* passages, have been so fully explored and made available in standard music of every grade, that necessarily the public taste has set itself to appreciate the pianoforte solo and the accompanied song exactly as they are written and performed. These are, after all, the highest forms of music which civilisation has yet enabled one or two performers to produce.

Yet, in regard to solo instrumentalisation, there is no doubt that a general hope exists for the discovery of a compromise between the piano and the organ or between the piano and the string band. Some inventors have aimed in the latter direction and others in the former; but no one has succeeded in really recommending his ideas to the public. Combined piano-violins and piano-organs have been shown at each of the great Exhibitions from the middle of the nineteenth century to its close. Several of these instruments have been devised and constructed with great ingenuity; and yet practically all of them have been received by the musical profession either with indifference or with positive ridicule.

The fact is that revolutionary sudden changes in musical instruments are rendered impossible owing to the near relationship which exists between each instrument and the general body of the music that is written for it. No one can divorce the two, which, as a factor in æsthetic progress, are really one and indivisible. Therefore, if any man invents a musical instrument which requires for its success the sudden evolution of a new race of composers writing for it, and a new type of educated public taste to hail these composers with delight, he is asking for a miracle and he will be disappointed.

What is wanted is not a new instrument, but an improved piano that shall at one and the same time correct, to some extent, the defects of the existing instrument, and leave still available all the brilliant effects which have been invented for it by a generation of musical geniuses. We want the sustained note, and yet we do not wish to lose the pretty turns and graceful devices by which the lack of it has been hidden, or atoned for, in the works of the masters. Therefore our sustained note must not be too aggressive. For a long time, indeed, it must partake of the very defects which it is intended ultimately to abolish.

In other words, we want to retain the percussion note with the dampers and with the loud and soft pedals, in fact, all the existing inventions for coaxing some of the notes to sustain themselves while others are cut short, as may be desired, and at the same time we have to add other and more effective means to assist the performer in achieving the same object.

The more or less complicated methods aiming at the prolongation of the residual effect of the percussion have apparently been very nearly exhausted. Some of the most modern pianos are really marvels of mechanical ingenuity applied to this purpose. We have now to look to something slightly resembling the principle of the violin or of the organ, in order to secure the additional *sostenuto* effect for which we are searching. Having to deal with a piano in practically its existing form, we obviously require to take special account of the fact that the note is begun by percussion, and that any attempt to bring a solid substance into contact with the wire while still vibrating, with the object of continuing its motion, is likely to produce more or less of a jarring effect.

The air-blast type of note-continuer for *sostenuto* effect therefore offers the most promising outlook for the improvement of the modern piano in the direction indicated. By directing a blast of air from a very thin nozzle on to the vibrating wire of a piano, the sound emitted may be very greatly intensified; and although naturally the decreasing amplitude of the vibration may in itself tend to create a *diminuendo*, yet it is possible to make up for this in some degree by causing the air-blast to increase in force, through the use of any suitable means, modified by an extra pedal as may be desired.

Delicate *pianissimo* effects, somewhat resembling those of the Eolian lyre, are produced by playing the notes with the air-blast alone, without the aid of percussion. But the louder *sostenuto* notes depend upon the added atmospheric resistance offered by a strong current of air to those movements of the wire which have been originally set up by percussion, and the fact that this resistance gives rise to a corresponding continuance of the motion. The prolongation of a note in this way is analogous to the continual swinging of an elastic switch in a stream of water, the current by its force producing a rhythmic movement.

When these Eolian effects, as applied to the pianoforte, have been carefully studied, many devices for controlling them will be brought forward. The main purpose, however, must be to connect the air-blast with the percussion apparatus in such a manner that, as soon as a key is depressed, the nozzle of that particular note in the air-blast is opened exactly at the same time that the wire is struck by the hammer, and it remains open as long as the note is held down. The movement of an extra pedal, however, has the effect of throwing the whole of the air-blast apparatus out of gear and reducing the piano to a percussion instrument, pure and simple.

It will be on the concert platform, no doubt, that this kind of improvement will find its first field of usefulness. Performers will require, in addition to their grand pianos, reservoirs of compressed air attachable by tubes to their instruments. In private houses hydraulic air-compressors will be found more convenient. When the piano has by some such means acquired the faculty of *singing* its notes, as well as of *ringing* them, its ascendency, as the finest instrument adapted to solo instrumentalism, will be assured.

The common domestic piano is rightly regarded by many people as being little better than an instrument of torture. One reason for this aversion is that, in the great majority of cases, the household instrument is not kept in tune. Probably it is not too much to say that the man who would invent a sound cottage piano which would remain in tune would do more for the improvement of the national taste in music than the largest and finest orchestra ever assembled. The constantly vitiated sense of hearing, which is brought about by the continual jangle of notes just a fractional part of a tone out of tune, is responsible for much of the distaste for good music which prevails among the people. When the domestic instrument is but imperfectly tuned, it is natural that those pieces should be preferred which suffer least by reason of the imperfection, and these, it need hardly be remarked, generally belong to the class of music which must be rated as essentially inferior, if not vulgar.

The device of winding a string round a peg and twisting it up on the latter in order to obtain tension for a vibrating note is thousands of years old. It was the method by which tension was imparted to some of the earliest harps and lyres of which history is cognisant; and it is still to be found to-day in the most elaborate and costly grand piano, with but few alterations affecting its principle of action. The pianoforte of the future will be kept in tune by more exact and scientific methods, attaining a certain balance between the thickness of the wire and the tension placed upon it by means of springs and weights.

Besides the ravages of the badly-tuned piano, much suffering is inflicted by the barbarous habit of permitting a sounding instrument to be used for mere mechanical exercises. The taste of the pupil is vitiated, and the nerves of other inmates of the house are subjected to a source of constant irritation when long series of notes, arranged merely as muscular exercises, and some of them violating almost every rule of musical form, are ground out hour after hour like coffee from a coffee-mill. The inconsistency of subjecting the musical ear and taste of a boy or girl to this process, and then expecting the child to develop an innate taste for the delicacies of form in melody and of the beauty of harmony, is almost as bad as would be that of asking a Chinese victim of foot-binding to walk easily and gracefully.

The use of the digitorium for promoting the mechanical portion of a musical education by the training of the fingers has already, to some slight extent, obviated the evils complained of. But this instrument is, as yet, only in its rudimentary stage of development. The dumb notes of the keyboard ought to be capable of emitting sounds by way of notice to the operator, in order to show when the rules have been broken. Thus, for instance, the impact caused by putting a key down should have the effect of driving a small weight upwards in the direction of a metal bar, the distance of which can be adjusted. Another bar, at a lower level, is also approached by a second weight, and the perfect degree of evenness in the touch is indicated by the fact that the lower bar should be made to emit a faint sound with every note, but the higher one not at all. The closer the bars the more difficult is the exercise, and remarkable evenness of touch can be acquired by a progressive training with such an instrument.

The organ has been wonderfully improved during the nineteenth century. Yet the decline of its popularity in comparison with the pianoforte may be accounted for on very rational grounds. While ardent organists still claim that the organ is the "King of Instruments" the public generally entertain a feeling that it is a deposed king. It remains for the organ-builders of the twentieth century to attack the problem of curing its defects by methods going more directly to the root of the difficulty than any hitherto attempted.

As contrasted with the pianoforte, the organ is extremely deficient in that power which the conductor of an orchestra loves to exercise—facility in accentuating and in subduing at will the work of each individual performer. For all practical purposes the ten fingers of a piano-player are the ten players in an orchestra; and, according to the force with which each finger strikes the note, is the prominence given to its effects. An air or a *motif* may be brought out with emphasis by one set of fingers, while the others are playing an accompaniment with all sorts of delicate gradations of softness and emphasis.

By multiplying the manuals, the organ-builder has endeavoured, with a certain degree of success, to make up for the unfortunate fact that the performer on his instrument possesses no similar facility in making it speak louder when he submits the note to extra pressure. One hand may be playing an air on one manual, while the second is engaged in the accompaniment on another; and the former may be connected with a louder stop, or with one of a more penetrating quality than the latter.

This device, together with an elaborate arrangement of swells and pedal-notes, has greatly enlarged the capacity of the organ for producing those choral effects which mainly depend upon gradations of volume. Yet the whole system, elaborate as it is, offers but a poor substitute for the marvellous range of individuality that may be expressed on the notes of the piano by instantaneous changes in the values ascribed to single notes. By the same action of his finger the pianist not only makes the note, but also gives its value; while the method of the organist is to neglect the element of finger-pressure and to rely upon other methods for imparting emphasis or softness to his work.

An organ that shall emit a louder or softer note, according to the force with which the key on the manual is depressed, will no doubt be one of the musical instruments of the twentieth century. Whether each key will be fitted with a resisting spring, or whether the lever will be constructed in such a way as to throw a weight to a higher or lower grade of position, according to the force with which it is struck, is a question which will depend upon the results of experiment. But the latter method is more in consonance with the conditions which have given to the piano its wonderful versatility, and it therefore seems the more probable solution of the two. Upon the vigour of the finger's impact will depend the height to which a valve is thrown, and this will determine the speed and volume of the air which is liberated to rush into the pipe and make the note.

The nineteenth century orchestra is a fearfully and wonderfully constructed agglomeration of ancient and modern instruments. Its merits are attested by the fine musical sense of the most experienced conductors, whose aim it has been so to balance the different instruments as to produce a tastefully-blended effect, while at the same time providing for solos and also for the rendering of parts in which a small number of performers may contribute to the unfolding of the composer's ideas. The orchestra cannot therefore be examined or discussed from a mechanical point of view, however much some of the instruments of which it is composed may be thought capable of improvement.

But the position of the conductor himself in the front of an orchestra is, from a purely artistic standpoint, highly anomalous. It is as if the prompter at the performance of a drama were to be seen taking the most conspicuous part and mixing among the actors upon the stage. If an orchestral piece be well played without the visible presence of a conductor, the sense of correct time reaches the audience naturally through the music itself; and any sort of gesticulations intended to mark it are under these conditions regarded as being out of place.

The foremost orchestral conductors of the day are evidently impressed with this unfitness of the mechanical marking of time by the wild waving of a stick or swaying of the body; and accordingly, however much they exert themselves at the rehearsal, they purposely subdue their motions during a public performance. The time is not far distant when the object of the conductor will be to guide his band without permitting his promptings to be perceived in any way by the audience.

For this purpose an "electric beat-indicator" will prove useful. Various proposals for its application have been put forward, and for different purposes several of them are obviously feasible. For instance, in one system the conductor sits in a place hidden from the audience and beats time on an electric contact-maker, which admits of his sending a special message to any particular performer whenever he desires to do so. The signal which marks the time may be given to each performer, either visually by a beater concealed within a small bell-shaped cavity affixed to his desk or to his electric light; or it may be conveyed by the sense of touch through a mechanical beater within a small metal weight placed on the floor and upon which he sets one of his feet.

The electric time-beater in the latter system thus taps the measure gently on the sole of the performer's foot, and special signals, as may be arranged, are sent to him by preconcerted combinations of taps. The absence of any distraction from the music itself will soon be gratefully felt by audiences, and the playing of a symphony in the twentieth century, in which the whole orchestra moves sympathetically in obedience to the "nerve-waves" of the electric current, will be the highest possible presentment of the musical art.

CHAPTER XIII
ART AND NEWS

The production of pictures for the million will be practically the highest achievement of the graphic art in the twentieth century. Many eminent painters do not at all relish the prospect, being strongly of opinion that when every branch of art becomes popular it will be vulgarised. This notion arises from a fallacy which has affected ideas during the nineteenth century in many matters besides art, the mistake of supposing that vulgar people all belong to one grade of society.

Yet every one who knows modern England, for instance, is perfectly aware that the highest standard of taste is only to be found in the elect of all classes of society. After the experience of the eighteenth century, surely it ought to have been recognised that the "upper ten thousand," when left to develop vulgarity in its true essence, can attain to a degree of perfection hardly possible in any other social grade. Is there in the whole range of pictorial art anything more irredeemably vulgar than a "State Portrait" by Sir Thomas Lawrence or one of his imitators?

It was under the prompting of a dread of the process of popularising art that so many eminent painters of the nineteenth century protested against the fashion set by Sir J. E. Millais when he sold such pictures as "Cherry Ripe" and "Bubbles," knowing they were intended for reproduction in very large numbers by mechanical means. From a somewhat similar motive a few of the leading artists of the nineteenth century for a time stood aloof from the movement for familiarising the people with at least the form, if not the colouring, of each notable picture of the year. From small and very unpretentious beginnings, the published pictorial notes of the Royal Academy and other exhibitions of the year have risen to most imposing proportions; and already there is some talk of attempting a few of the best from each year's production in colours.

Half-tone zinco and similar processes have brought down the expenses entailed by reproductions in colour-work, so as to render an undertaking of this kind much more feasible than it was in the middle of the last half-century. "Cherry Ripe" cost five thousand pounds to reproduce, by the laborious processes of printing not only each colour, but almost every different shade of each colour from a different surface.

In the "three-colour-zinco" process of reproduction only three printings are required, each colour with all its delicate gradations of shade being fully provided for by a single engraved block. When machines of great precision have been finally perfected for admitting of the successive blocks being printed from on paper run from the reel without any handling, a revolution will be brought about not only in artistic printing, but even in the conditions of studio work upon which the artist depends for success.

First, the pictorial notes of the year will be brought out in colour; and as competition for the right of reproduction increases, the artists who have painted the most suitable and most popular pictures will find that they can get more remuneration for copyright than they can for the pictures themselves. This has already been the case in regard to a very limited number of pictures; but the exception of the past will be the rule of the future, at least as regards those pictures which possess any special merits at all.

More thought will therefore be required as the motive or basis of each subject; and historical pictures will come more into favour, the affected simplicity and mental emptiness of the *plein air* school being discarded in favour of a style which shall speak more directly to the people, and stir more deeply both their mental and their emotional natures.

The artist and the printer must then confer. They can no longer afford to work in the future with such disregard of each other's ideas and methods as they have done in the past. It was at one time the custom among painters almost to despise the "black and white man" who drew for the Press in any shape or form; but that piece of affectation has nearly been destroyed by the general ridicule with which it is now received, and by the knowledge that there are already, at the end of the nineteenth century, just as many men of talent working by methods suitable for reproduction, as there are painters who confine their attention to palette, canvas and brush.

The printer will now advance a step further, and will invoke the services of the painter himself, even prescribing certain methods by which the Press may be enabled to reproduce the work of the artist more faithfully than would otherwise be possible.

Transparency painting will no doubt be one of these methods. The artist will paint on a set of sheets of transparent celluloid or glass, mounted in frames of wood and hinged so that they can, for purposes of observation, be put aside and yet brought back to their original positions quite accurately.

Each different transparent sheet will be intended for one pure colour, the only pigments used being of the most transparent description obtainable.

The picture may thus be built up by successive additions and alterations, not all put upon one surface, but constituting a number of "monochromes," superimposed one upon the other. When finished, each of these one-colour transparencies can then be reproduced by photo-mechanical means for multi-colour printing in the press.

By what are known as the photographic "interruption" processes, a kind of converse method has achieved a certain degree of success. A landscape or a picture is photographed several times from exactly the same position, but on each occasion it is taken through a screen of a different coloured glass, which is intended for the purpose of intercepting all the rays of light, except those of one particular tint. Coloured prints in transparent gelatine or other suitable medium are then made from the various negatives, each in its appropriate tint; and when all are placed together and viewed through transmitted light, the effect of the picture, with all its colours combined, is fairly well produced. More serviceable from the artistic point of view will be the method according to which the artist makes his picture by transmitted light, but the finished printed product is seen on paper, because this latter lends itself to the finest work of the artistic printer.

The principal branch of the work of the photographer must continue to be portraiture. He cannot greatly reduce the cost of getting a really good negative, because so much hand-labour is required for the task of "retouching"; but he can give, perhaps, a hundred prints for the price which he now charges for a dozen, and make money by the enterprise. It has already been proved that there is no necessity for using expensive salts of gold, silver or platinum in order to secure the most artistic prints; and, as a matter of fact, some of the finest art work in the photography of the past quarter of a century has been accomplished with the cheapest of materials, such as gelatine, glue and lampblack.

Pigmented gelatine is, without doubt, the coming medium for photographic prints, and the methods of making them must approximate more and more closely to those of the typographic printer. By producing a "photo-relief" in gelatine—sensitised with bichromate of potash, and afterwards exposed first to the sun and then to the action of water—an impression in plastic material can be secured, from which, with the use of warm, thin, pigmented gelatine, a hundred copies or more can be printed off in a few minutes.

The very general introduction of such a process has naturally been delayed owing to the extra trouble involved in the first methods which were suggested for applying it, and also, no doubt, on account of the recent fashion for platinotype and bromide of silver prints. But as soon as more convenient details for the making of pigmented gelatine prints have been elaborated, the cheapness of the material and the wonderful variety of the art shades and tints in which photographs can be executed will give the gelatine processes an advantage in the competition which it will be hopeless for other methods to challenge.

The daily newspapers of a few years hence will be vividly illustrated with photographic pictures of the personages and the events of the day. The gelatine photo-relief, already alluded to, will no doubt afford the basis of the principal processes by which this will be effected. Hitherto the chief drawback has been the difficulty of imparting a suitable grain to the printing blocks made from these reliefs; but this has been practically overcome by the use of sheets of metallic foil previously impressed with the form of a finely-engraved tint-block. The actual printing surface, of course, consists of an electrotype or stereotype taken from this metallic-grained photographic face.

For "high-art" printing on fine paper with the more expensive kinds of ink, the half-tone zinco processes will doubtless maintain their supremacy and gradually diminish the area within which lithographic printing is required. In the case of newspaper work, however, where haste in getting ready for the press is necessarily the prime consideration, the flat and very slightly-indented surface of the zinco block is found to be unsuited to the requirements. Flat blocks, which require careful "overlaying" on the machine, waste too much time for daily news work. Without going into technical details it may be surmised in general terms that in the near future almost every newspaper will contain, each day, one or more photo-illustrations of events of the previous day or of the news which has come to hand from a distance.

Type-setting by hand is, for newspaper purposes, being so rapidly superseded, that only in the smaller towns and villages can it remain for even a few years longer. But in the machines by which this revolution has been effected, finality has been by no means reached. Every line of matter which appears in any modern daily newspaper has to pass through two processes of stereotyping before it makes a beginning to effect its final work of printing upon paper.

First, there is the stereotyping or casting of the line in its position in the type-setting machine after the matrices have been ranged in position by the application of the fingers to the various keys; and, secondly, when all the lines have been placed together to make a page, it is necessary to take an impression of them upon *papier mâché*, or what is technically called "flong," and then to dry it and make the full cast from it curved and ready for placing on the cylinder of the printing machine. The delay occasioned by the need for drying the wet flong is such a serious matter—particularly to evening newspapers requiring many editions during the afternoon—that several dry methods have been tried with greater or less success.

But there is really no need for more than one casting process. In the twentieth century machine the matrices will be replaced by permanent type from which, when ranged in the line, an impression will be made by hard pressure on a small bar of soft metal or plastic material. All the impressed bars having been set together in a casting box having the necessary curvature, the final stereo plate for printing from will be taken at once by pouring melted metal on the combined bars.

An appreciable saving, both in time and in money, will also be effected by applying the principle of the perforated strip of paper or cardboard to the purpose of operating the machine by which the necessary letters are caused to range themselves in the required order. Machines similar to typewriters will be employed for perforating the strips of paper and for printing, at the same time, in ordinary letters the matter just as if it were being typewritten.

The corrections can then be made by cutting off those pieces of the strips which are wrong and inserting corrected pieces in their places. No initial "justification" to the space required to make a line is needed in this system. The strips, however, are put through the setting machine, and, as they make the reading matter by the impression of bars as already described, they are divided into lines automatically.

Large numbers of newspapers will in future be sold from "penny-in-the-slot" machines. The system to be adopted for this particular purpose will doubtless differ in some important respects from that which has been successful in the vending of small articles such as sweetmeats and cigarettes. The newspapers may be hung on light bars within the machine, these being supported at the end by a carefully-adjusted cross piece, which, on the insertion of a penny in the slot, moves just sufficiently to permit the end of one bar with its newspaper to drop, and to precipitate the latter on to a table forming the front of the machine. When the full complement of newspapers has been exhausted the slot is automatically closed.

Some of the newspapers of the twentieth century will be given away gratis, and will be, for the most part, owned by the principal advertisers. This is the direction in which journalistic property is now tending, and at any juncture steps might be taken, in one or other of the great centres of newspaper enterprise, which would precipitate the ultimate movement.

Hardly any one who buys a half-penny paper to-day imagines for a moment that there is any actual profit on the article.

It is understood on all hands that the advertisers keep the newspapers going and that the arrangement is mutually beneficial. Not that either party can dictate to the other in matters outside of its own province. The effect is simply to permit the great public to purchase its news practically for the price of the paper and ink on which it is conveyed; the condition being that the said public will permit its eyes to be greeted with certain announcements placed in juxtaposition to the news and comments.

Sooner or later, therefore, the idea will occur to some of the leading advertisers to form a syndicate and give to the people a small broadsheet containing briefly the daily narrative. The ponderous newspapers of the latter end of the nineteenth century—filled full of enough of linotype matter to occupy more than the whole day of the subscriber in their perusal—will be to a large extent dispensed with; and the new art of journalism will consist in saying things as briefly—not as lengthily—as possible.

CHAPTER XIV
INVENTION AND COLLECTIVISM

The ownership of machinery and of all the varied appliances in the evolution of which inventive genius is exercised is a matter which, strictly speaking, does not belong to the domain of this work. Nevertheless, in endeavouring to forecast the progress of invention during the twentieth century, it is necessary to take count of the risks involved in the inauguration of any public and social economical systems which might tend to stifle freedom of thought and to discourage the efforts of those who have suggestions of industrial improvements to make.

It is plain that those economic forces which prevent the inventor from having his ideas tested must to that extent retard the progress of industrial improvement. Thousands of men, who imagine that they possess the inventive talent in a highly developed degree, are either crack-brained enthusiasts or else utterly unpractical men whose services would never be worth anything at all in the work of attacking difficult mechanical problems. It is in the task of discriminating between this class and the true inventors that many industrial organizers fail. Any economic system which offers inducements to the directors of industrial enterprises to shirk the onerous, and at times very irksome, duty of sifting out the good from the bad must stand condemned not only on account of its wastefulness, but by reason of its baneful effects in the discouragement of inventive genius.

Considerations of this kind lead to the conclusion that during the twentieth century the spread of collectivist or socialistic ideas, and the adoption of methods of State and municipal control of production and transport may have an important bearing upon the progress of civilisation through the adoption of new inventions. Many thinking men and women of the present generation are inclined to believe *the* twentieth century invention *par excellence* will be the bringing of all the machinery of production, transport and exchange under the official control of persons appointed by the State or by the municipality, and therefore amenable to the vote of the people. Projects of collectivism are in the air, and high hopes are entertained that the twentieth century will be far more distinctively marked by the revolution which it will witness in the social and industrial organisation of the people than in the improvements effected in the mechanical and other means for extending man's powers over natural forces.

The average official naturally wishes to retain his billet. That is the main motive which governs nearly all his official acts; and in the treatment which he usually accords to the inventor he shows this anxiety perhaps more clearly than in any other class of the actions of his administration. He wants to make no mistakes, but whether he ever scores a distinct and decided success is comparatively a matter of indifference to him. So long as he does not give a handle to his enemies to be used against him, he is fairly contented to go on from year to year in a humdrum style.

Even a man of fine feeling and progressive ideas soon experiences the numbing effects of the routine life after he has been a few years in office. He knows that he will be judged rather on the negative than on the positive principle, that is to say, for the things which it is accounted he ought not to have done rather than for the more enterprising good things which it is admitted he may have done.

Now any one who undertakes to encourage invention must necessarily make mistakes. He may indeed know that one case of brilliant success will make up for half a dozen comparative failures; but he reckons that at any rate the blanks in the chances which he is taking will numerically exceed the prizes. An official, however, will not dare to draw blanks. Better for him to draw nothing at all. He must therefore turn his back upon the inventor and approve of nothing which has not been shown to be a great success elsewhere.

This means that the socialised and municipalised enterprises must always lag behind those depending upon private effort; and the country which imposes disabilities on the latter must, for a time at least, lose its lead in the industrial race. This is what happened to England, as contrasted with the United States, when, under the influence of enthusiasm for future municipalisation, the British Legislature laid heavy penalties upon those who should venture to instal electric trams in the United Kingdom.

The American manufacturers and tramway companies, in their keen competition with one another and perfect freedom to compete on even terms with horse traction, soon took the lead in all matters pertaining to electric traction, and the British public, at the close of the nineteenth century, have had to witness the humiliating spectacle of their own public authorities being forced to import electrical apparatus, and even steam-engines applicable to dynamos used for tramway purposes, from the other side of the Atlantic!

The lesson thus enforced will not in the end be missed, although it may require a considerable time to be fully understood. Officialism is a foe to inventive progress; and whether it exists under a regime of collectivism or under one of autocracy, it must paralyse industrial enterprise to that extent, thus rendering the country which has adopted it liable to be outstripped by its competitors.

The true friend of inventive progress is generally the rising competitor in a busy hive of industry where the difficulties of securing a profitable footing are very considerable. Such a man is ever on the watch for an opportunity to gain some leverage by which he may raise himself to a level with older-established or richer competitors. If he be a good employer his workmen enter into the spirit of the competition, feeling that promotion will follow on any services they may render. They may perhaps possess the inventive talent themselves, or they may do even greater services by recognising it in others and co-operating in their work. It is thus that successful inventions are usually started on their useful careers.

It is therefore upon private enterprise that the principal onus of advancing the inventions which will contribute to the progress of the human race in the twentieth century must necessarily fall. The type of man who will cheerfully work *pro bono publico*, with just as much ardour as he would exhibit when labouring to advance his own interests, may already be found here and there in civilised communities at existing stages of development; but it is not sufficiently numerous to enable the world to dispense with the powerful stimulus of competition.

Just as a superior type of machinery can be elaborated during the course of a single century, there is no doubt that—mainly through the use of improved appliances for lessening the amount of brute force which man needs to exert in his daily avocations—the nervous organisations of the men and women constituting the rank and file during the latter part of the twentieth century will be immensely improved in sensitiveness. A corresponding advance will then take place in the capacity for collectivism. But a human being of the high class demanded for the carrying out of any scheme of State socialism must be bred by a slow improvement during successive generations. A hundred years do not constitute a long period of time in the process of the organic evolution of the human race, and, as Tennyson declared,

> We are far from the noon of man—
> There is time for the race to grow.

Yet the public advantages of collectivist activities in certain particular directions cannot for a moment be denied. Much waste and heavy loss are entailed by the duplication of works of general utility by rival owners, each of them, perhaps, only half utilising the full capacities of his machinery or of the other plant upon which capital has been expended.

Moreover, as soon as companies have become so large that their managers and other officials are brought into no closer personal relations with the shareholders than the town clerks, engineers, and surveyors of cities, and the departmental heads of State bureaus are associated with the voters and ratepayers, the systems of private and of collective ownership begin to stand much more nearly on a par as regards the non-encouragement which they offer to inventiveness.

One of the greatest discoveries of the twentieth century, therefore, will be the adoption of a *via media* which will admit of the progressiveness of private ownership in promoting industrial inventions, combined with the political progressiveness of collectivism. One direction in which an important factor assisting in the solution of this problem is to be expected is in the removal of the causes which tend to make public officials so timid and unprogressive.

So long as a mere temporary outcry about the apparent non-success of some adopted improvement—whose real value perhaps cannot be proved unless by the exercise of patience—may result in the dismissal or in the disrating of the official who has recommended it, just so long will all those who are called upon to act as guides to public enterprises be compelled to stick to the most conservative lines in the exercise of their duties. More assurance of permanence in positions of public administration is needed.

The man upon whose shoulders rests the responsibility of adopting, or of condemning, new proposals brought before him, ostensibly in the interests of the public welfare, ought to be regarded as being called upon to carry out *quasi*-judicial functions; and his tenure of office, and his claim to a pension after a busy career, ought not to depend upon the chances of the evanescent politics of the day. If a man has proved, by his close and successful application to the study of his profession—as evinced in the tests which he has passed as a youth and during his subsequent career in subordinate positions—that he is really a lover of hard work, and imbued with conscientious devotion to duty, he may generally be trusted, when he has attained to a position of superintendence, to do his utmost in the interests of the public whom he serves. This is the theory upon which the appointment of a judge in almost any English-speaking community is understood to be made; and, although failures in its application may occur now and then, there is no doubt whatever that on the average of cases it works out well in practice.

If private manufacturers, whose success in life depends upon their appreciation of talent and inventiveness, could be assured that in dealing with public officials they would be brought into contact with men of the standing indicated, instead of being confronted so frequently with the demand for commissions and other kinds of solatium on account of the risks undertaken in recommending anything new, they would soon largely modify their distrust of what is known as collectivism. It is the duty of the public whose servant an official is, rather than of the private manufacturer, to insure him against the danger of losing his position on account of any possible mistake in the exercise of his judgment.

In short, the day is not far distant when the men upon whom devolves the responsibility of examining into, and reporting upon, the claims of those who profess to have made important industrial improvements will be looked upon as exercising judicial functions of the very highest type. When the important reforms arising from this recognition have been introduced, the forces of collectivism will cease to range themselves on the side of stolid conservatism in industry, as they undoubtedly have done in the nineteenth century even while they inconsistently professed to advance the cause of progress politically.

The inventor, who in the early part of the nineteenth century was generally denounced as a public enemy, will, in the latter part of the twentieth century, be hailed as a benefactor to the community, because he will be judged by the ultimate, rather than by the immediate, effects of his work, and because it will be the duty of the public authorities to see to it that the dislocation of one industry incidental the promotion of another by any invention does not, on the whole, operate to throw people out of employment, but, on the contrary, gives more constant work and better wages to all. But the slow progress of the fundamental traits of human nature will retard the attainment of this goal. The world has a long distance to travel in the uphill road of industrial and social improvement before it can succeed in obtaining a really true view of the part fulfilled by inventive genius in contributing to human happiness.

www.ingramcontent.com/pod-product-compliance
Lightning Source LLC
LaVergne TN
LVHW042156190726
843493LV00006B/1701